SECTION - I

INTRODUCTORY

Vedic Mathematics Basics

Published by :
Lotus Press Publishers & Distributors

Vedic Mathematics Basics

Dr. S.K. Kapoor
Ved Rattan

4735/22, Prakash Deep Building
Ansari Road, Darya Ganj,
New Delhi - 110002

Lotus Press : Publishers & Distributors
Unit No. 220, 2nd Floor, 4735/22, Prakash Deep Building,
Ansari Road, Darya Ganj, New Delhi- 110002
Ph.: 41325510, 98118-38000
• E-mail : lotuspress1984@gmail.com
www.lotuspress.co.in

Vedic Mathematics Basics

ISBN: 81-8382-045-X

Printed & Published by : **Lotus Press Publisher & Distributors,** New Delhi-02

INNER VOICE FOREWORD

— FROM A SAINT

Hear your inner voice and be free from the outside limitations. Within is the seat of the Lord and it is here where we have to be face to face with our selves. This origin source reservoir is to be tapped, and this appears to be the urge and intensity of present attempt to reach at 'Vedic mathematics basics' in five steps. These five steps would prepare for the phase and stage to be in communication with the SELF.

The SELF is TRANSCENDENTAL. The seat of the SELF within *Triloki* is the Creator's space. Self established at the seat of Creator's space ensures carrying the Being along the transcendental carriers. It is the bliss of this ensured supply of transcendental carriers which intensifies the urge of the transcending mind to glimpse the inner folds of the transcendental worlds manifesting within Creator's space.

The whole efforts and beginning for it, as such is to be to attain transition from Triloki manifesting as outside mundane macro state world as first state of consciousness for the intelligence to the next transcendental micro state world of the intelligence lively within Creator's space as origin source reservoir of Triloki. Our ghan / body cube, as such, is to be chased for its origin source reservoir within it lively as Chatar mukhi / four head Creator.

Really blissful is the learning path of this chase range. It is transcendentally blessed. Let sadkhas permit their transcending mind to fulfill with bliss of the transcendental worlds / Panch Mukhi / five head lord sustaining the Creator's space.

Lively within cavity of heart: 25-11-2005	In search of ***JYOTI*** manifesting as **NAD**

■

PREFACE

Urge to know ancient wisdom expressed as Vedic knowledge inherently imples acceptance for chase of it the way these systems permit. This as such would mean that the sadkhas for the chase of Vedic systems have to proceed along the established paths of *Sankhiya Nishshta* and *Yoga nishta* which not only run parallel to each but are also complementary and supplementary of each other.

The sankhiya nishtha, on its ultimate analysis amounts to availing the artifices of numbers, for coordination arrangement of organisation format, while on the other hand, the Yoga nishta avails the dimensional formats, for regulation of transcendence and ascendance of mind through the manifested creations. The **TRANS**, also means, the number ahead, and it is this way, the transcending mind ultimately avails the artifices of numbers to sequentially glimpse the ultimate core of the transcendental worlds.

The par-excellence feature of Vedic systems is that these approach the whole range of knowledge as a single Discipline, and as such, on ultimate analysis the organisation of Discipline of knowledge avails the dimensional frames parallel to the artifices of numbers. This, this way, shall be taking us to the artifices of numbers, and the dimensional frames as the source processing tools in terms of which even the self referral feature of transcending mind chasing its own transcending path can also be worked out intellectually.

It is this feature which deserves to be chased for the basics, and here in the present attempt, the urge being to learn and teach Vedic mathematics on its basics, as such, at this phase and stage of exposing young minds during Middle school learning, The focus is to be upon the centre of the cube / origin of 3-space being of the format of hyper cube-4 and as such seat of creator's space.

This focus naturally aims to eye the goals of attainments for the young minds for perfection of their intelligence by having transition from the set up of the Cube to transcend ahead for glimpsing the set up of hyper cube-4 along with its transcendental base of the format of hyper cube-5.

Though, the learning all about hyper cubes 4 and 5 is to be the content focus of high school classes, and ahead there to would be the subject focus of the formats of hyper cubes-6 and 7 during higher secondary classes, but the initiations for the foundations of appropriate mental state for those formats are to be ensured during schooling of middle school stage.

This way the learning range of this phase and stage of schooling during classes 5 to 8 is to be centred around the set up of the cube with focus upon its centre /

origin of 3-space as seat of creator's space (4-space) as of format of hyper cube -4. Accordingly the teaching and learning of this range of schooling is phased as five books namely, book-1The teaching of Vedic mathematics, book-2 Learning Vedic mathematics on first principles, book-3 Vedic mathematics basics, book-4 Vedic mathematics skills and book-5 Vedic geometry course, for perfection of intelligence.

These books, in their integrated approach of chase of the range for set up of the cube / 3-space with focus upon its center / origin of 3-space, as source reservoir for the features of the set up, is meant for exposure of young minds to acquire perfection of intelligence while learning basis of basics on first principles along the concrete format of a cube.

The first principles for approaching basis of basics of Vedic mathematics on solid formats are being supplied by the Ganita Sutras, particularly as the ordering principle of Ganita Sutra-1 which makes it as the source Sutra, and as such 'cube' and 'Ganita Sutra-1' are to ever remain within the eye sight of the learners of this phase and stage of schooling.

The aim and goal of learning being the chase of basis of basics on first principles to ensure perfection of intelligence for the young minds, as such heavy responsibiliy lies upon the Vedic mathematics teachers and parents to make the learning really blissful throughout the duration of chase.

The whole range of enlighened souls begining with Maharishi Valmiki and Maharishi Ved Vyas to Maharishi Dayanand and Swami Bharti Krsna TirthaJi Maharaj to Shri Pad Baba and Maharishi Mahesh Yogi, this learning path is being made blissfully lively, and the enlightened teachers are to continue the linege.

With these inspired background, present humble attempt has been made to reach at these books and it is hoped that these may provide required exposure for the young minds with the help of the enlightened teachers, who shall be making their instructions free of technicalities and the knowlede content to flow blissfully within the head s and hearts of the students.

Mr. Kenneth R. Williams and Mr. R. P. Cosmic Kapoor deserve my heartful thanks for remaining intellectually with me during these days of my reaching at these books. Mr. Rohtash Kapoor deserves full appreciation for conducting Vedic geometry course.

Mr. Deepak Girdhar deserves all appreciation for his nice graphics and the computer work.

Credit for nice publication goes to Lotus Press.

The inner voice forward from a saint gives me the satisfaction and confidence as that these efforts may be fruitful for the cause of basics to be brought home to the young minds.

Vedic home, **Author**
Rohini, Delhi
25-11-2005

■

CONTENTS

1

INTRODUCTORY

1.1 INTRODUCTORY

1. '***Sankhiya***' (numbers, artifices of numbers; knowledge path) and '***Yoga***' (unison, unison through transcendence of geometric formats; actions path) are two established paths ('***Nishtha***') of ancient wisdom (Gita 3.3).
2. Both, Sankhiya and Yoga, run parallel to each other and attain the same end fruit. Both these paths are to be glimpsed as complementary and supplementary of each other. And of those who so glimpse are really glimpsing (Gita 5.4 & 5.5).
3. Both, Sankhiya and Yoga nishtha, values lead to glimpsing of 'atman' / soul display domain during '***dhyan***' (meditation) (Gita 13.24).
4. As such the basic values of truth coming within comprehension of mind are to be as the glimpsing domain of transcending mind which with its transcendence of artifices of numbers and geometric formats shall be leading to the end fruit of ambrosia of display expression domain of atman / soul.

5. Therefore, the most basic step of learning is the meditation, that is, to permit one's mind to transcend. The initial preparatory steps are in the form of a seat of suitable setting to have a comfortable sitting so that the mind has transcendence in its natural path without any external obstructions in the form of noise etc. The choice of posture is to be for a prolonged sitting of trans, for which the avoidance of strain is to be had by having the setting of body, neck, head and eye sight following the nose line till its tip end. (Gita 6.11 to 14).

1.2 URGE TO KNOW

1. *Sadhkas*, having
   - (a) having an intensified urge,
   - (b) to know,
   - (c) and chase,
     - the transcendental glimpse of *Vedic* mathematics

   shall,
   - (a) sit comfortably and,
   - (b) permit the mind to transcend,
     - to glimpse the transcendental world,
     - as transcendental phenomena,

   and have,
   - (a) self validation of,
   - (b) the way and path,
     - of transcendental glimpse of *Vedic* mathematics.

2. Between two sittings of trans, the *Sadhkas,* shall,
   (a) go for the intellectual exercise of to be through the scriptures,
   (b) and also share the experiential bliss of transcendental world with fellow *Sadhkas,*
   (c) and have self validation of experiences, as well as to be sure about the perfection of intelligence gained through transcendence and through intellectual exercise.

(Glimpses of Vedic mathematics: Chapter-3 page 85)

1.3 OLD FORMAT TO NEW FORMAT

1. *Sadhkas,* having,
   (a) an intensified urge to know,
   (b) and to chase,
      (i) the transcendental glimpse of *Vedic* Mathematics,
      (ii) of its way and path of chasing the transition,
      (iii) from its old format of organization of knowledge to its new format of organization of knowledge,

   shall,
   (a) sit comfortably, and transcend time and again,
   (b) and glimpse,
      (i) the transcendental world,
      (ii) of its way and path of unfolding and folding back of its inner folds of its own,

(c) and during time gaps of two such sittings of trans, shall also

(i) go through the intellectual exercise of going through the scriptures for perfecting of intelligence,

(ii) and also to share the experiential bliss with other fellow *Sadhkas*, for insuring self-validation of the transcendental comprehension.

2. This exercise,

(a) of chasing the transcending mind is very delicate exercise, and as such it shall be,

(b) approached in small sequential steps, and for it,

(a) the start with state is,

(i) the waking state of consciousness,

(ii) and from it, take off to be had,

(b) by transcendence from waking state of consciousness to dream state of consciousness,

(i) for perfection of intelligence,

(ii) and for chiseling of third eye,

with ambrosia of bliss of experiential glimpse of the inner folds of the transcendental world unfolding and folding back of its own.

3. The scriptures are full of preservation of experiential bliss shared by *Sadhkas* having glimpse the transcendental world, with fellow *Sadhkas*, as that

(a) the waking state of consciousness is the state

of linear order of our *Vishwa* (*Triloky*(⊞)/three space of linear dimensional order)

(b) while the dream state of consciousness is the state of spatial order of creator's the space (4-Space(⌑) of spatial dimensional order).

4. *Vedic* mathematics,

(a) in its unique way,

(b) chases the transition,

(i) from the waking state of consciousness to dream state of consciousness,

- as mind transcending,
- for glimpse of the transcendental world,
- of its inner folds,
- unfolding and folding back of its own,

(ii) and the transcendental phenomena,

- happening,
- fulfilling the mind,
- having glimpsed,
- the inner most fold of the transcendental world,
- with ambrosia of bliss of the inner most fold of the transcendental world,

(c) and with it,

(i) the mind ascends,

(ii) with perfected intelligence,

(d) and in the process,

- additional eye having been chiseled for the,
- head at the joint of the pairs of eyes.

5. The *Sadhkas*, having an intensified urge,
   (a) to reaches this chase of transcendental glimpse of *Vedic* Mathematics,
   (b) from the old format of waking state of consciousness to the new format of the dream state of consciousness,

   shall sit comfortably and permit the mind to transcend time and again

   (i) and to chase the transcending mind,
   (ii) step by step, as small sequential leaps,
   (iii) starting from the linear order of 3-Space (▣) as waking state of consciousness,
   (iv) to the spatial order of 4-Space(⌑) as dream state of consciousness,

   and having complete transition from old format of linear order of 3-Space(▣) to new format of spatial order of 4-Space(⌑).

   (Glimpses of Vedic mathematics: Chapter-3 pages105-107)

1.4 NUMBERS AND GEOMETRIC FORMATS

1. First state of consciousness is designated as ***'jagrit awastha'*** / waking state of consciousness. The parallel geometric format is of Vishwa / Jagat / Triloki / 3-space / cube.
2. As such the intensity of urge at this initial stage is to be centered around and is to be to fully melt

and to completely chase the set up of a 'cube' as representative regular body of 3-space.

3. The transcending mind as it would be through the set up of cube / 3-space as solid boundary of Creator's space / 4-space with seat at origin of 3-space / center of cube, it shall be creating its own intelligence field and in the process shall be perfecting its intelligence for glimpsing the domain of Creator's space / 4-space / hyper cube 4.

1.5 AMBROSIA OF ATTAINMENTS

1. It would be the fruit of being through the process of learning and transcendence through the set up of waking state of consciousness / set up of Triloki / Jagat / 3-space / Cube / solid boundary enveloping within itself the Creator's domain / 4-space / hyper cube 4 format.
2. It may be taken as the ambrosia of bliss of the transcendence from the linear order (1-space in the role of dimension of 3-space domain) of macro-state of existence phenomenon to spatial order (2-space in the role of dimension of 4-space domain) of microstate of existence phenomenon.

■

For further studies

'Vedic Geometry' be refrred for the basic Vedic geometrical concepts, and for special symbols, as well as for the pre dominant features of real four, five, and six spaces.

'Glimpses of Vedic mathematics' may be referred for the glimpsing domains of transcending mind and the connected aspects. ■

2

BASICS SOURCE RESERVOIR

Waking state consciousness leads to '**cube**' as the 'basics source reservoir'. Let us have a fresh look at the 'cube' (A) and know more and more about it.

2.1 OF SYNTHETIC ENVELOPE

'Cube' is a set up of 'synthetic envelope'. The constituents of this 'envelope' are (1) Eight corner points (2) Twelve edges and (3) Six surfaces. This being constituted of these geometric entities, the same as such is designated 'geometric envelope' of 26 entities.

2.2 ARTIFICE TWENTY SIX

The artifice twenty-six is a very rich artifice. The ancient wisdom has availed it for chase of 26 basic elements. Of these, the 26th, as ultimate, is chased, in terms of first 25 of them, designated as '25 elements of *sankhiya darshan*'. Literally, 'sankhiya' means 'number'. '***Sankhiya***' is one of the two basic paths (*nishtha*), namely 'sankhiya and yoga'.

There are 26 basic meters (*chandas*). ***Chandas*** literally means 'filters'. It filters the values flow from the '**Ultimate'** as well as the flow back into the

'Ultimate'. This chase is a quarter by quarter chase, and as such, this meters / measures range comes to be of 26 x 4 = 104 meters for complete coverage of the sky in between ***Prithavi*** / Earth and **Suriya** / *Sun*.

2.3 VALUES FLOW CHASE

The values flow chase of chandas / meters for its *vridhi* / increase and gunna / features avails 'pairing of counts', as a basic operation.

This basic 'operation' of 'pairing' of 'counts' for 'flow' of 'values' along 26 meters / filters as 26 basic elements, as such deserves to be chased for reaching at 'the set up of a cube'.

2.4 PAIRING OF COUNTS

The pairing of counts 1 to 26, when chased shall be, firstly fixing the starting as well as the end step of the chase as beginning with count '1' and reaching up till the count '26'.

Under the process of the operation which has been taken to the 'quarters', it makes simultaneous and outward expansions, as that '1' takes to ' ½ ' for itself to be a product of 'pairing' of ' ½ ' with itself. Simultaneously 'under the operation of pairing, '1' also takes to '2'.

This, this way takes us from '26' to '13', for elements themselves being products of 'pairing', and further the pairing of 26 with itself takes to '52'.

Parallel to this chase is the chase of alphabets, of which the alphabet of 52 letters covers up till artifice 13 for its quarters and also takes to artifice 104 for availing pairing for the letters.

This as such makes not only the 'letters' themselves

as a product of pairing operation, but also constitutes meters / filters measures with pairing of 'letters' as 'counts'.

2.5 BASICS' SOURCES AND SOURCE RESERVOIR

This way the basics' sources and source reservoir emerged to be

(1) 'Set up of a cube'

(2) 'Waking state of consciousness'

(3) '26 Basic elements'

(4) '26 Basic meters of 104 quarters meters range'

(5) '52 letters alphabet'

(6) '26 letters alphabet'

(7) 'Artifices 1 to 26'

(8) 'Pairing of counts'

(9) 'Artifice 13'

2.6 ARTIFICE 13

The cube avails 12 edges, which may help fix it in a static position. Its motion shall be manifesting the path of motion as its additional, that is, 13th edge.

The static position of a cube be taken as its fixation as within a 3-space. Within 4-space, the availability of an additional dimension shall be providing a degree of freedom of motion along the fourth dimension, which as such, as a path of motion of a cube shall be manifesting as 13th edge of a cube and shall be making it a hyper cube.

The spatial order of 4-space shall be making the 13th edge of the cube as being of a spatial order, and thereby, the whole set up would get transformed as of the value of artifice 26 being of the order 2 x 13.

2.7 TWENTY SIX LETTERS ALPHABET

The English alphabet is 26 letters alphabet (A to Z). For values flow chase, the formats of these letters shall be filtering as artifices 1 to 26 spreading in the sequence and order of alphabet letters A to Z. This shall be associating artifices values of range 1 to 26 to the alphabet letters range A to Z in that sequence and order and thereby (A, 1), (B, 2) and so on (Z, 26) would get paired.

The pairing of counts operation shall be accepting A as 1, B as 2 and so on Z as 26. These may be formally taken by way of definition as that number value format (In short NVF) of 'A', that is, NVF (A) =1, NVF (B) =2, NVF (C) =3, NVF (D) =4, NVF (E) =5, NVF (F) =6, NVF (G) =7, NVF (H) =8, NVF (I) =9, NVF (J) =10, NVF (K) =11, NVF (L) =12, NVF (M) =13, NVF (N) =14, NVF (O) =15, NVF (P) =16, NVF (Q) =17, NVF (R) =18, NVF (S) =19, NVF (T) =20, NVF (U) =21, NVF (V) =22, NVF (W) =23, NVF (X) =24, NVF (Y) =25, NVF (Z) =26.

This further shall be yielding number value formats for the 'WORDS' as formulations of settings of individual letters which in terms of their counts values shall be through pairing of counts leading to number value formats for the WORDS FORMULATIONS. Illustratively NVF (WORD)= NVF (W) +NVF (O) + NVF (R) +NVF (D)=23+15+18+4=60.

Likewise the groups of words shall be attaining number value formats through pairing of counts operations running through the letters of all the words of group. Illustratively NVF (GROUP OF WORDS) = NVF

(GROUP) + NVF (OF) + NVF (WORLDS)= (7+18+15+21+16) +(15+6)+ (23+!5+18+4+19)= 67+21+79=167.

■

For Further Studies

'Vedic mathematics decodes SPACE BOOK' may be referred for further studies about the applied values of the basic pairing operation of 26 basic elements along artifices 1 to 26 and parallel words formulations availing number value formats of same generic. This book is based on the postulates and axioms of interesting features. It accepts the emergence and existence of black substance along with the light fulfilling the space. The pairing operaion accepts di monad format of which one part is dedicated to the black substance and the other part is dedicated to the white substance. The working with one part only as half dimension is the distinguishing skill of the system which helps to reach at the linear order as a spatial case of the system based in the special order of the creator's space (4-space). The glimpsing transcendental world reached at in the book is its feature which is bound to absorb the curious minds.

■

3

ORDERING PRINCIPLE AT THE BASE

3.1 INTRODUCTORY

1. Processing chase basics are provided by Ganita Sutras and Upsutras.
2. There are sixteen Sutras and thirteen Upsutras in all and of these, the first Sutra supplies the ordering principle, while the first Upsutra supplies the symmetry chase rule.
3. These two principles, namely (1) ordering principle of Sutra-1 and (2) symmetry chase rule of Upsutra-1, may be taken as the start with basics which shall be running through the whole range of processing processes of the Ganita Sutras system for their chase of transcendence through artifices of numbers and geometric formats.

3.2 GANITA SUTRA-1

एकाधिकेन पूर्वेण

Ekadhiken Purvena

One more than before

1. Ganita Sutra-1 reads 'एकाधिकेन पूर्वेण' and sounds

as '**Ekadhiken Purvena**'.

2. The simple English rendering for its functional rule, may be: '**one more than before**'.

3.3 AFFINE AND SEQUENTIAL SETTINGS

3. At the base of this functional rule 'one more than before' is the basic ordering principle which fulfills the '**affine setting**' into '**sequential ordering**'.
4. The '**affine setting**', may be taken as it is for '**any display expression**'. Illustratively all trees as these are without any order taking from 'one tree to other'.
5. The sequential ordering shall be taking us from 'one tree to the other'.
6. The infinite points fulfilling infinite surface, in its affine setting, when is embedded with a pair of Cartesian axes supplies a pair of coordinates for each point of the surface, and thereby, a pair of sequential orders of the pair of coordinates may get associated with the whole range of points of the surface, which shall be converting it into a Cartesian plane.

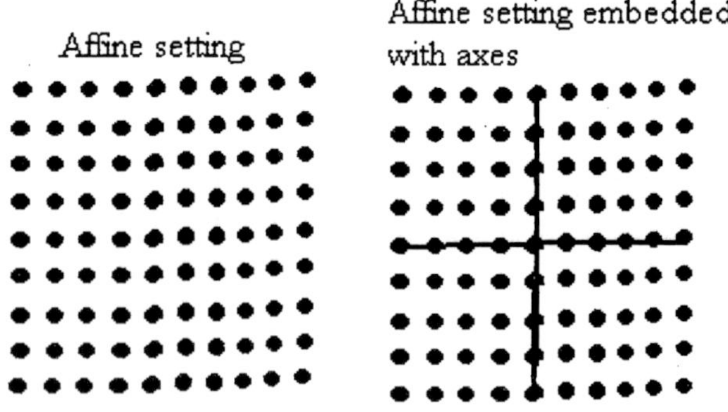

7. Now if the original pair of axes are transformed

into another setting of 45 degree with the original setting, the infinite surface points shall be constituting a second Cartesian plane along the same affine setting of the surface.

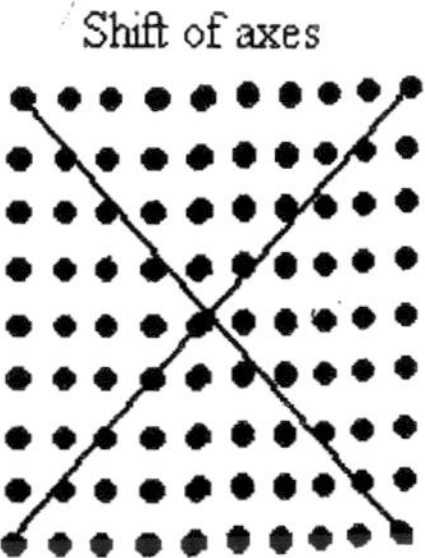

8. It is this feature of affine setting remaining undisturbed while it simultaneously getting different sequential orderings, which deserves to be chased.

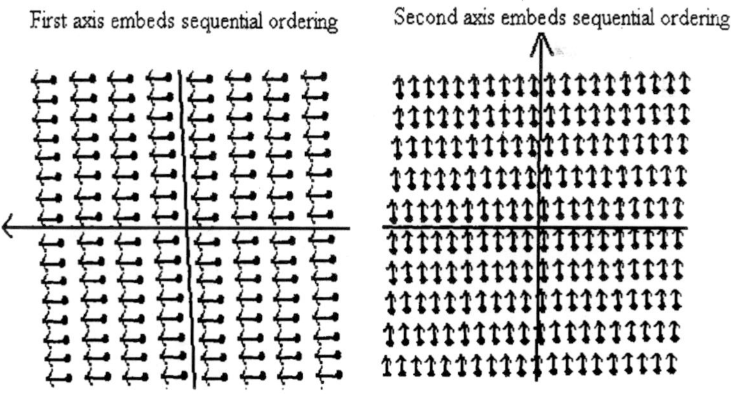

With transformation of axes, therebeing transformation of sequential ordering

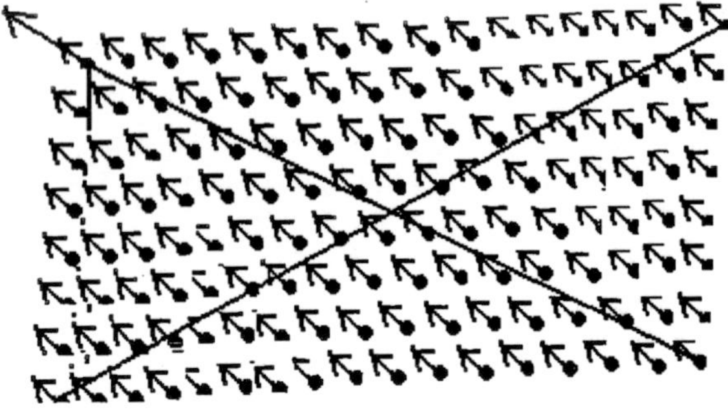

3.4 CHASE OF TRANSITION FROM AFFINE TO SEQUENTIAL SETTING

9. 'One way to chase this phenomenon of a shift from affine setting to it gaining sequential ordering is to proceed with the rule of 'one more than before'.
10. It is this chase on first principles which takes us to the Ganita Sutras system starting with its very first Sutra and the very first Upsutra of this system.

3.5 UPSUTRA-1

11. Before this chase is followed, it would be appropriate to first acquits ourselves with the text of Ganita Upsutra-1 and its functional rule.

आनुरूप्येण ANURUPYENA

' To follow the form as it is framed' /

As proportionate (symmetric order)

4

FINITE AND INFINITE COUNTS

4.1 INTRODUCTORY

1. The ordering rule 'one more than before', adds one 'count', to the count already reached at.
2. Taking the count reach at being count '5', the ordering rule, shall with addition of one more count be taking the counting to count '6'.
3. With it, there is, not going to be termination of the counting at any count, as under the rule, there would follow another count being one count more than the count already reached at.
4. This, this way makes any given count being achievable in finite number of steps, while the process as a promise of 'no termination up till any finite steps' and as such it is to apply infinitely.
5. It is this unique feature, which makes attainment of any count with finite steps only, while the process continuous ahead and ahead, and takes to 'infinity'.

6 This feature, as such, very help us comprehend as to while 'infinite' is 'in-finite'.

4.2 FINITE AND INFINITE ARE OF DIFFERENT FEATURES

1. The infinite is the end up till which the finite (counts) reach at.
2. The infinitely extended line transform into the circumference of a circle and this way sustains the infinite linear steps along finite circumference.
3. This feature of line transforming as circumference of a circle and absorbing infinite linear steps along finite circumference, deserves to be chased as to how 'infinite' is 'in-finite'.
4. This transition and transformation for the line into circumference deserves to be chased, firstly as that here there has been a transition and transformation for the format beneath that of 'line' as of 'one space' into that of 'circumference' which is sustained by a circle, a 2-space set up / body.
5. It is this transition and transformation for setting of the points of a line, from that of 1-space to that of 2-space, which makes all the wonderful characteristics of this feature of 'infinite' being 'in-finite'.
6. The circumference of a circle being 'a line, though curved' is to contribute 'zero area' as the line is only thin and is devoid of 'width'.
7. The circle within circumference is a surface of positive value of area (may it be $0^2 = 0 \times 0$), and as such it is of distinct generic than that of line order of circumference which accepts only 'linear measure' and as such how so over big it may be, because of absence of second axis, it is to always remain devoid of area value.

8. The circumference because of absence of 'width', and is to be 'devoid of area', as such it is 'cipher', while even a point circle shall be of 'zero' area.
9. The distinction between CIPHER and ZERO is significantly there though may be at very 'fine' comprehension level.
10. One feature of distinction between 'CIPHER' and 'ZERO' is that 'ZERO' is a 'UNIT' and is of identity as is a point of a plane, while 'CIPHER' is devoid of any identity as simply it is not having membership of the exclusive club of 'areas'. How so ever long the line / circumference may be, it cannot earn 'area' and hence cannot expect a membership of exclusive club of areas.
11. The circumference is to always remain at boundary and to be 'cipher' for the surface-domain of the circle.
12. Circumference / line while claiming its domain status in its own right as linear domain, it shall be like wise excluding membership to its boundary points and getting them tagged as 'ciphers', while the points within the domain (along the line) to be of 'zero length'.
13. This domain–boundary distinction for the setups of the dimensional bodies, in general, and closed interval, square, cube and hyper cubes in particular as representative regular dimensional bodies deserves to be chased.
14. Beginning for this chase may be had with the set up of a cube, as it is a representative regular body of 3-space and the same is to be of the order of our Triloki / Vishwa / Jagat / Space of our existence of waking state consciousness order. ■

5

RECAPITULATION AND SUM UP

GEOMETRIC ENVELOPE OF CUBE AND ARTIFICE 26

Synthetic geometric envelope of cube of 26 entities

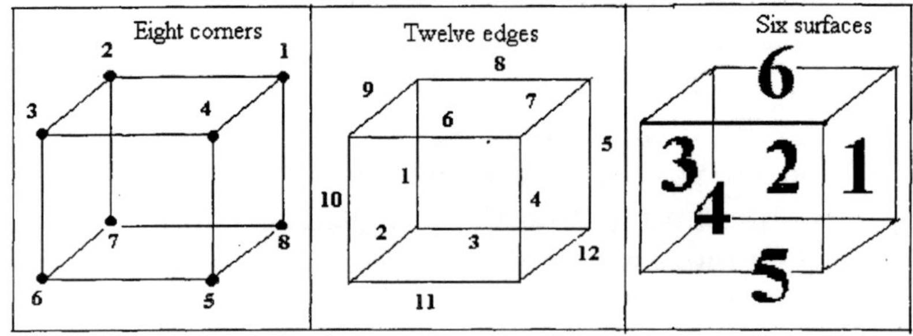

5.1 INTRODUCTORY

1. Let us have a fresh look at the set up of a 'cube' and see as that the domain / volume of the cube is enveloped within a fully stitched set up of eight corner points, twelve edges and six surfaces.
2. Therefore before reaching at the features of the domain / volume part of the cube, it may be appropriate to have first of all the comprehension

of this synthetic set up of geometric envelope of the cube.

5.2 GEOMETRIC ENVELOPE

3. One distinguishing feature of this geometric envelope of cube comes to be that it avails artifice 26 for its synthetic set up with internal arrangement as of artifices 8, 12 and 6.

5.3 ARTIFICE 26

4. The artifice 26 is of many distinguishing features which makes it a very rich artifice.
5. Scriptures preserve the enlightenment as that there are 26 basic elements, and parallel to it there are 26 basic meters, and thereby, there are 26 x 4 =104 range of meters quarters, which being of the features of meters extend the range of meters to be of measure 104 for covering the sky between Earth and Sun.
6. These 26 basic elements and corresponding meters accept coordination arrangements in terms of artifices of numbers as counts accepting pairing for synthetic setups of the quarters range to be of the order as being double of the number of the letters of the alphabet.
7. This coordination arrangement of meters makes part of the whole to be of the features of the whole, and with it the whole space features are filtered into its each representative regular body as being envelops synthesized in terms of the features of the basic elements.
8. This as such makes simultaneous available the inward as well as outward expansions following

the coordination arrangement of artifices of numbers of the format 1 x 1 = 2 / 1 x ½ .

9. The double value along one axis and half value along the second axis retains the spatial unit undisturbed, and this feature, when availed by the system, it acquires self referral features as is the case of Ganita Sutras system.
10. Further when this self referral feature becomes available within each of the quarter (of a plane / spatial order). The origin gets isolated giving way for transcendence.
11. It is this transcendence feature which makes the system to be of for excellence features of sequential manifestation and transcendence ad-infinitem bringing 'infinite' within 'finite'.
12. As such the basics chase is to be in terms of the set up of a cube, and parallel to its synthetic envelope of 26 elements admitting pairing along artifices of numbers 1 to 26, the chase would get facilitated as a realistic language of an alphabet of 26 letters yielding words formulations with individual letters of the words contributing equal to their number value formats, which are to be of the range 1 to 26.
13. As such, (1) 26 basic elements, (2) 26 artifices need of sankhiya system for chase of the ultimate as 26th element in terms of 25 elements, (3) 26 basic meters, (4) 26 entities constituting synthetic geometric envelope of cube and (5) 26 letters English alphabet emerge to be the source reservoir of features to be worked out for basics in terms of the first principles supplied by the Vedic

mathematics system of Ganita Sutras and Upsutras.

■

For further studies

'Vedic mathematics Ganita Sutras', may be referred for deep insight into the Ganita Sutras system. This book chases the organisation format as well as the functional format of the Ganita Sutras system accepting Ganita Sutra-1 as the source Sutra and in terms of its ordering principle the whole range of the system from Ganita Sutra-1 itself to Ganita Sutra-16 and from Ganita Upsutra-1 to Ganita Upsutra-13, whole range unfolds itself parallel to the composition format of Ganita Sutra-1.

Once, one is through this volume, one may have answers for many questions which may be striking curious minds, particularly regarding the potentialities of this system, its postulates, source and the base at which these Sutras are to be approaced.

■

SECTION - II

SET UP OF A CUBE

1
INTRODUCTORY

1.1 INTERVAL, SQUARE AND CUBE

Interval	Square	Cube
——	□	(cube)
1-space body	2-space body	3-space body

1. Let us have a fresh look at the cube.
2. Edge of the cube is a track of a moving point. It is of a format of an interval (1-space body).
3. The surface plate of the cube is a track of a line / interval / edge. It is of a format of a square (2-space body).
4. The volume / solid / domain of the cube is a track of a surface plate. It is a format of 3-space body.

1.2 $A^N: 2N B^{N-1}$, N=1, 2, 3, 4

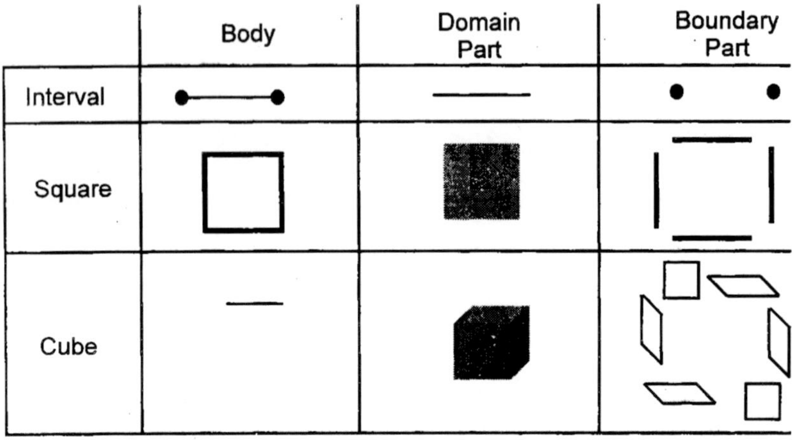

	Body	Domain Part	Boundary Part
Interval			
Square			
Cube			

1. Let us again have a fresh look at the set ups of cube, square and interval.
2. Interval is a set up between a pair of end points. It may be expressed in terms of domain (length / A^1) and boundary (end points / $2B^0$) ratio formulation A^1: $2B^0$.
3. Square is a set up between a double pair of boundary lines. It may be expressed in terms of domain (area / A^2) and boundary (end lines / $4B^1$) ratio formulation A^2: $4B^1$.
4. Cube is a set up between a triple pair of boundary surface plates. It may be expressed in terms of domain (volume / A^3) and boundary (end surfaces plates / $6B^2$) ratio formulation A^3: $6B^2$.
5. These three domain boundary formulations for interval, square and cube may be expressed as a common formulation A^N: $2NB^{N-1}$, N=1, 2, 3.

6. It would be a blissful exercise to project the sequence term ahead of 'interval, square and cube' and to have its domain-boundary ratio values in terms of the above common formulation for its value as N=4.
7. The sequential term ahead of 'interval square and cube' is designated as hyper cube 4 as of domain boundary ratio being A^4: 8 B^3.
8. The symbolic representation for this hyper space (4-space) body with solid boundary of eight components may be as follows

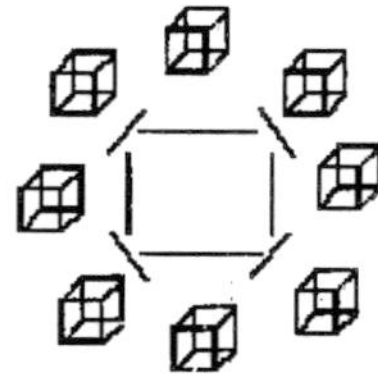

1.3 $(A+2)^N$, N=1, 2, 3

1. Let us have a fresh look at the set ups of interval, square and a cube.
2. It may be observed as that the set up of an interval is of constituents (1) length / A^1 and (2) pair of end points / 2 A^0. These together may be expressed as $(A^1+2\ A^0)^1$.

3. Further it may be observed as that the set up of the square as of constituents (1) area / A^2 (2) double pair of boundary lines / $4A^1$ and (3) double pair of end points / 4 A^0. These together may be expressed as $(A^1+2\ A^0)^2 = A^2 + 4\ A^1 + 4\ A^0$.
4. Still further it may be observed as that the set up of the cube as of constituents (1) Volume / A^3 (2) area of surface plates / $6A^2$ (3) length of edges / $12A^1$ and (4) corner points / 8 A^0. These together may be expressed as $(A^1+2\ A^0)^3 = A^3 + 6A^2 + 12\ A^1 + 8\ A^0$.
5. These individual expressions of set ups of interval, square and cube in terms of their constituents, may be jointly expressed as single formulation being.

   $(A^1+2\ A^0)^N$, N=1, 2, 3.

1.4 (3, 3-1), (3, 3-2)

1. Let us have a fresh look at the set up of the cube and observe as that its volume is of third power (volume / A^3) and its surface boundary is of second power (area / A^2).
2. This domain-boundary inter-relationship, as 3-space and 2-space respectively, and also as third power and second power respectively, may be expressed as artifice 3 and artifice 2 respectively. It without loss of generality may be expressed as (3, 3-1) feature of the set up of the cube.
3. Further the simultaneous expression for domain / volume of the cube as A^3 and for dimension / axis of the cube as A^1 shall be taking us to the expression (3, 3-2) for this another feature of the set up of the cube.

4. Both these features as an expression (3, 3-1, 3-2) may be taken as that in the set up of a cube its volume / domain accepts third power expression (A^3), boundary / surface area accepts second-degree expression (A^2) and the dimension / axis measure accepts single degree expression (A^1).

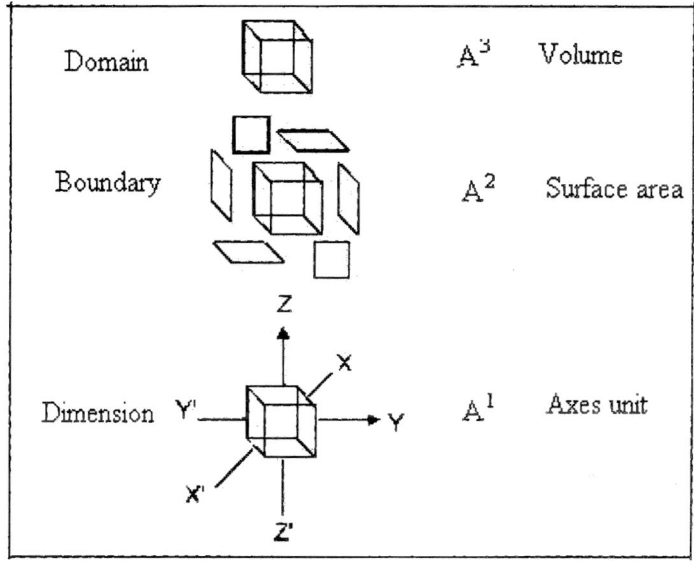

5. The general expression of such coordination for domain, boundary and dimension for the set up of a representative regular body of N-space may be accepted for the present as (N-Space as domain, N-1 Space as boundary, N-2 Space as dimension / axis).

1.5 13TH EDGE

1. Let us again have a fresh look at the set up of a cube.
2. It may be observed as that this set up is framed in

terms of twelve edges.

3. It may be accepted, by definition, as a position in rest for the set up as framed within its twelve edges frame.
4. This rest position, may be taken as a fixation of the set up within 3-space.

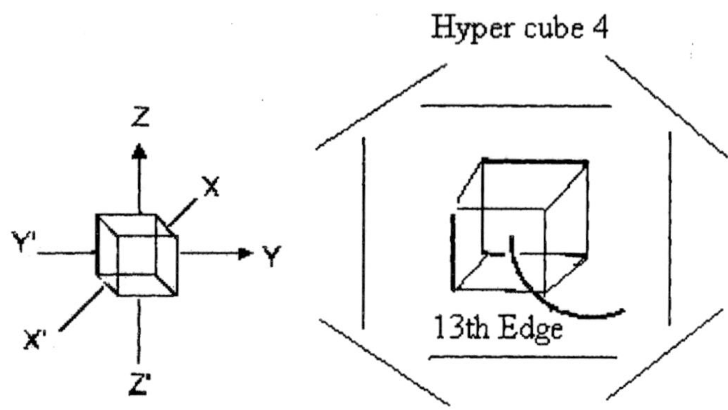

With availability of a degree of freedom of motion along an additional dimension within 4-space, its track (path) of motion shall be taken as manifesting an additional edge for the cube.

■

For further studies

One shall refer to the scriptures about the distinguishing feature of waking state reality at the dream state reality as these are to be respectively of the formats of 3-space / cube and 4-space / hyper cube 4 format. ■

2
GEOMETRIC ENVELOPE

2.1 PAIRING OF COUNTS

1. It be taken an axiom of chandas-meters / filters measures as that each distinct constituent to accept a distinct 'count', and that the values flow through the constituents permits chase as 'pairing of counts'.

6-surfaces $6A^2$	12 edges $12A^1$	8 corners $8A^0$

2. The chase of geometric envelope of cube shall be taking us to artifice 26 as the end value for the counts values flow through all the constituents manifesting this synthetic envelope for the cube.

2.2 ARTIFICE 26

1. The artifice 26 covers the values flow through the constituents of geometric envelope of the cube.
2. The artifices 1 to 26 split 50 factors as follows

No.	Factors	Total	Grand Total	No.	Factors	Total	Grand Total
1	1	1	1	14	2 × 7	2	23
2	2	1	2	15	3 × 5	2	25
3	3	1	3	16	2 × 2 × 2 × 2	4	29
4	2 × 2	2	5	17	17	1	30
5	5	1	6	18	2 × 3 × 3	3	33
6	2 × 3	2	8	19	19	1	34
7	7	1	9	20	2 × 2 × 5	3	37
8	2 × 2 × 2	3	12	21	3 × 7	2	39
9	3 × 3	2	14	22	2 × 11	2	41
10	2 × 5	2	16	23	23	1	42
11	11	1	17	24	2 × 2 × 2 × 3	4	46
12	2 × 2 × 3	3	20	25	5 × 5	2	48
13	13	1	21	26	2 × 13	2	50

3. The artifice 26 accepts 50 factors, which is of the order of number value format for 'void' which

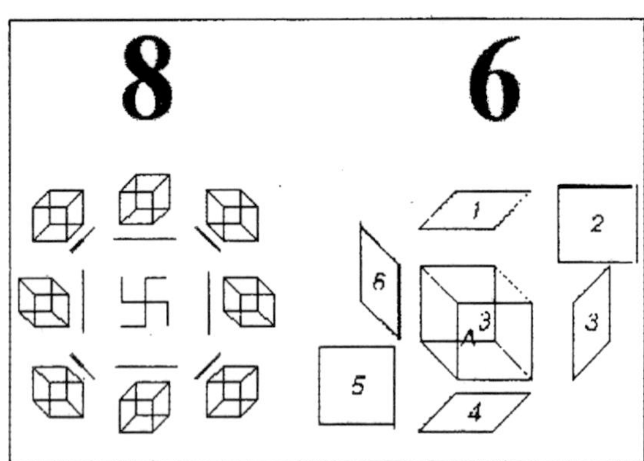

proceeds as well as transcends through all manifested values, and as such the 'real void' is 'para-graph' and is of the order of 'a creation'; NVF (PARAGRAPH) = NVF (REAL VOID) = NVF (A CREATION) = 86, which with its value 6 at unit / linear order placement is of the order of 6 boundary components of linear order 3-space, and ahead the spatial order placement value 8 being of the order of 8 boundary components of spatial order 4-space.

2.3 ARTIFICE 26 TO ARTIFICE 31

1. Let us again have a fresh look at the set up of the cube and observe as that the synthetic set up of the boundary as well as of the domain together with the dimensional axis at the center yield value of artifice 31 with a split up as 8 corner points, 12 edges, 6 surface plates, 1 volume, 3 axes and 1 center (origin).

Volume A^3	6-surfaces $6A^2$	12 edges $12A^1$	8 corners $8A^0$
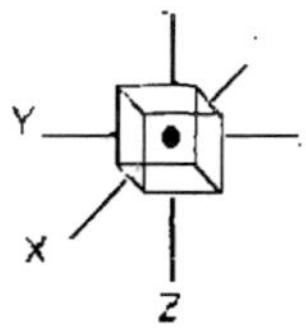 Y X Z	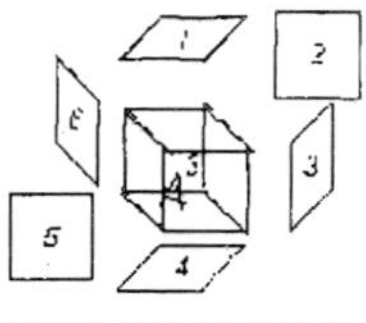 1 2 3 4 5 6	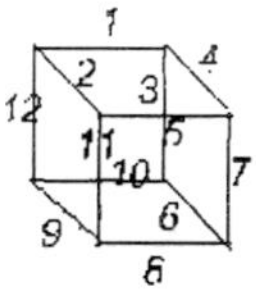 1 2 3 4 5 6 7 8 9 10 11 12	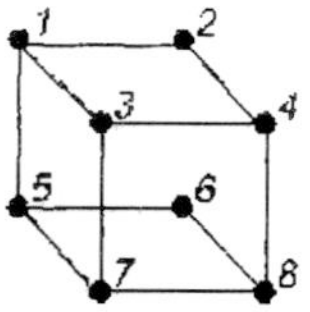 1 2 3 4 5 6 7 8

2. NVF (CUBE) = 31.

2.4 VOLUME

1. Volume of the cube is expression of 3-space content as domain of the cube.

2. It is an expression of third power (A^3).

3. It has fixation in terms of a three dimensional frame.
4. It accepts a geometric envelope of the order of artifice 26.
5. It accepts a solid grid for retention of solid format for its each granule.

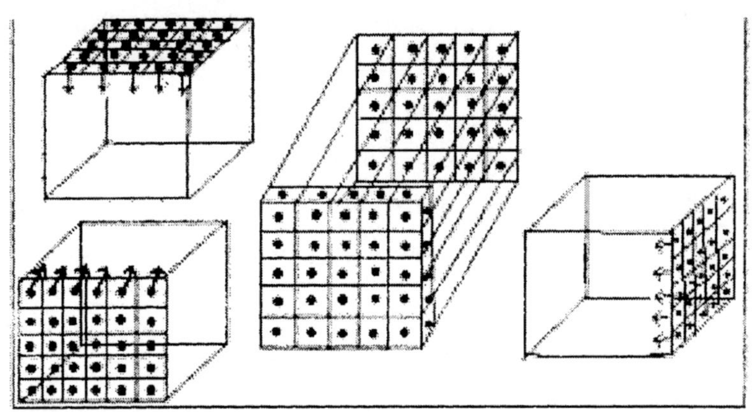

2.5 3-SPACE CONTENT

1. It be taken by way of definition as that volume of the cube is expression of the feature of 3-space content.
2. The feature of 3-space content as of volume expression permits its chase as solid grid cage for its granules.
3. In other words, be taken as that the 3-space content manifests as solid grid cages for its granules.
4. The availability of this expression format for each of the granule of 3-space content makes it to be its basic feature.

2.6 LIQUEFIED CONTENT

1. One of the characteristics of the basic feature of 3-space content emerges to be of it acquiring even zero value along any or any pair or for all of its axes.
2. This characteristics of the basic feature may be accepted as a basic characteristic of the basic feature.
3. This basic characteristic of basic feature of 3-space content may, without loss of generality may be designated and approached as 'basic of basics' of 3-space'.
4. It is in terms of this basic of basics that the values flow of set up of a cube may be chased for its synthetic boundary as well as for its domain and even for the center of the cube.
5. It is this 'basic of basics' characteristics-feature of 3-space which is to help to have a chase for the space content to be in a 'liquefied state' with one of the axes remaining at a dormant state, and even may in a limiting situation to be at zero value.
6. The range of values for the axis measure may be from zero to infinity, and as such, its limitations to different finite values, shall be manifesting a whole range of formats for the liquid states of contents of whole range of dimensional spaces.
7. As such, this characteristic feature of 'basic of basics' of set up of 3-space, deserves to be comprehended and chased as a sequence of liquefied content formats.

2.7 CONTENT FLOW

1. One way to look at the cube is that its boundary is constituted by grid zones.
2. The other way to look at the cube is a solid grid domain within twelve-edged frame.
3. The grid zones at the boundary of the cube shall be taken as permitting in flow of linear order through its centers into the domain of the cube which in the process of inflow from the centers of its boundary grid zones manifesting a solid order for the domain.
4. The twelve edged frame for the solid grid domain in terms of its coordination at eight corner points regulates flow as split streams of seven steps ranges which ultimately manifest as seven versions of a cube as representative bodies of seven geometries of 3-space.
5. Both these features of the set up of cube deserve to be chased to comprehend the flow of content of 3-space availing different formats which in the process not only manifest a synthetic envelope for sustaining 3-space content lump but also for sustaining the content flow by following the edges coordination and spatial grids together constituting the solid grids for the granules of 3-space content together expressing as volume of the cube.
6. It is the hyper cube format at center of cube / origin of 3-space which sustains the entire set up of a cube up till its zero value. The collapse of all the eight corner points along with the domain of zero value at center / origin with domain of cube

Seven geometries of 3-space with seven versions of the cube

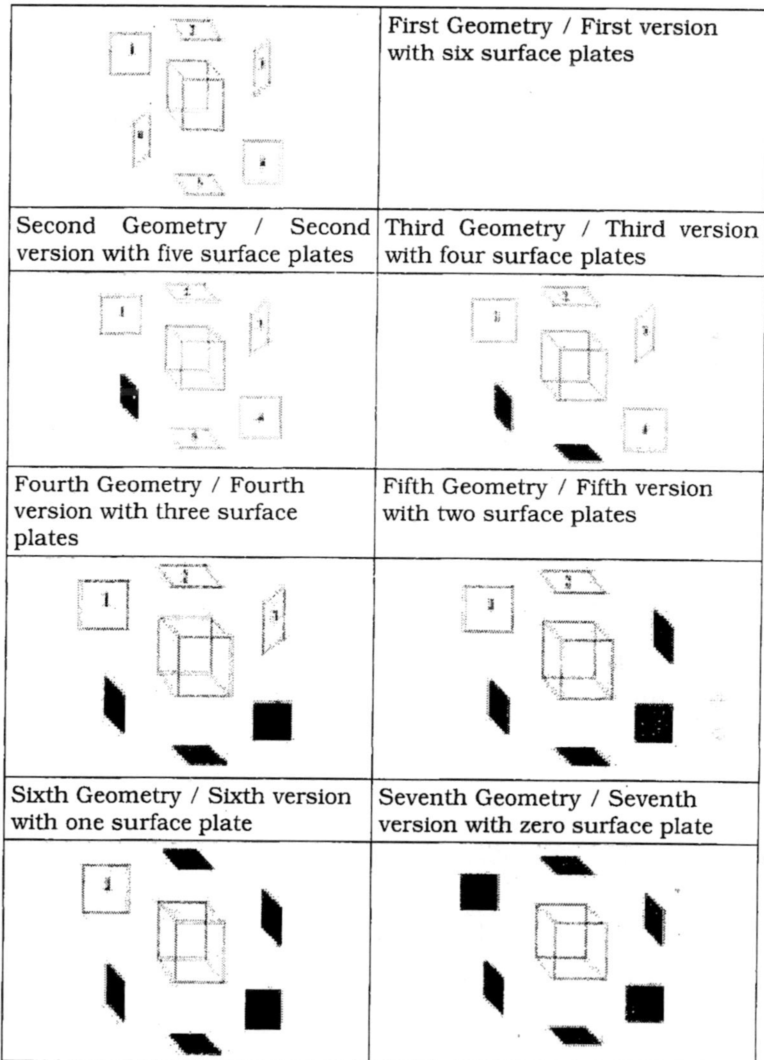

resting at the domain of hyper cube 4 and corners with 3 dimensional frames intact though at zero value but being of solid set up resting at the seats of eight solid boundary components of hyper cube 4.

7. It would be a blissful exercise to bring all the eight

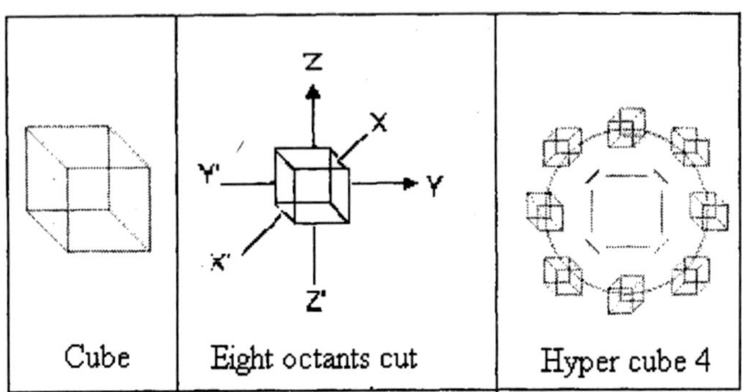

Cube	Eight octants cut	Hyper cube 4

octants nearer and nearer to the origin an parallel to it the eight cubes of a cube with values tending to the limiting zero value resulting into collapse of all the eight corner points with three dimensional frame along with the cube domain simultaneously resting at the format of hyper cube 4.

8. Simultaneously the reverse exercise of the zero value cube resting at the format of hyper cube 4 evolving into positive value cube with a three dimensional frame at the center of cube / origin of 3-space superimposed upon center of hyper cube 4 / origin of 4-space as a seat of solid order 5-space. The teachers shall help the students to be successfully through this blissful exercise.

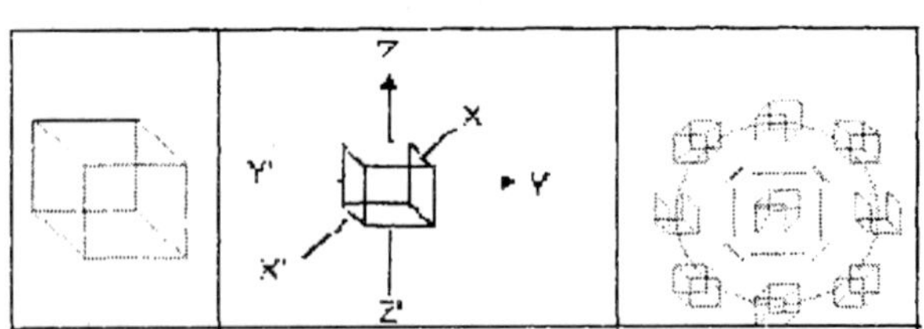

■

3

FLOW ALONG EDGES

3.1 FLOW ALONG EDGES

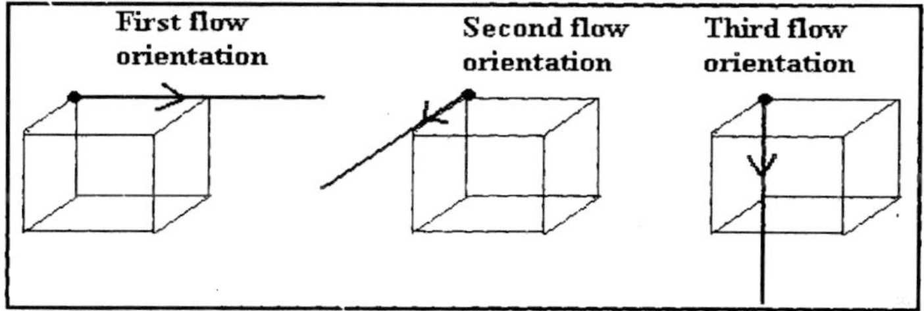

1. One shall see that three edges meet at each corner point of the cube.
2. One may view it as a feature of this set up as that from each corner of a cube their may be three distinct flow orientations along three distinct edges.
3. It would be taken as an exercise to carry forward along the orientation of a given edge to connect all the eight corner points.
4. Here, one may be confronted with a pair of options at the second step, as is being illustrated here under.

Step 1. To proceed along the first orientation and reach the other end of the edge at second corner of the Cube.

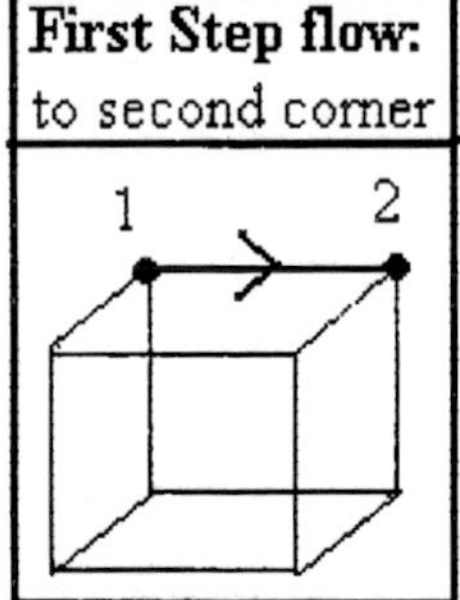

Step 2. At the second corner, the flow to have two options as under:

Step3. Either of the options, while reaching third corner, shall be again confronted with a pair of options. The same be followed individually. And, at next step as well there to follow a pair of options, but with that, the rest of the flow is to have only single flow option to connect remaining all the corner points.

5. It be taken, as an exercise to workout all the options for the original first orientation flow from the first corner.
6. Further it also be taken as an exercise to workout the flow options for the second and third flow orientations from the first corner.
7. Still further, it also be taken as an exercise to workout the similar flow options for all the eight corners of the cube.
8. Still further, as the flow orientation takes from first corner to the 8th corner in seven steps availing coordination arrangement of seven edges in a sequence, and as each such flow may be taken in reverse orientation, as starting with the 8th corner and reaching up till the first corner, as such the question would stand posed for an answer as to how many pairs of such orientations are accepted

by the set up of the cube. Before actual chase and working out of all the orientation flow coordination of all the 8th corner at a time, one may attempt the poser as to, if the total pairs of orientations taking from first corner to eighth corner, and in reverse orientation from eighth corner to first corner, to be precisely 8 x 3 =24.

9. One may pause and have a fresh look at the set up of the cube and to chase and comprehend on the format of the flow orientation connecting a pair of corner points in terms of a connecting edge, as that the edge, as a line is a flow of a moving corner / point.
10. One may attempt definition of a line as a flow track of a moving point.
11. One may attempt chase path for a flow track along the format of a line in a pair of ways parallel to the pair of the orientations of a line.
12. One may attempt to define a direction for the other end corner point in terms of the orientation of the edge leading from first corner point to the second corner point.

3.2 SPLIT STREAMS

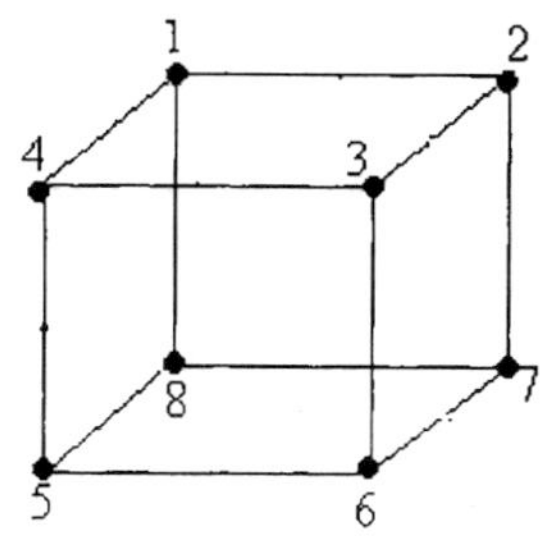

1. The students shall chase the possible number of ways all the eight corner points may be coordinated through edges of the cube by maintaining the orientation and without visiting any of a corner points more than once.
2. Let us fix our starting point to be corner '1'.
3. Let us fix our orientation for flow at the initial stage of corner '1' to be from corner '1' to corner '2'.

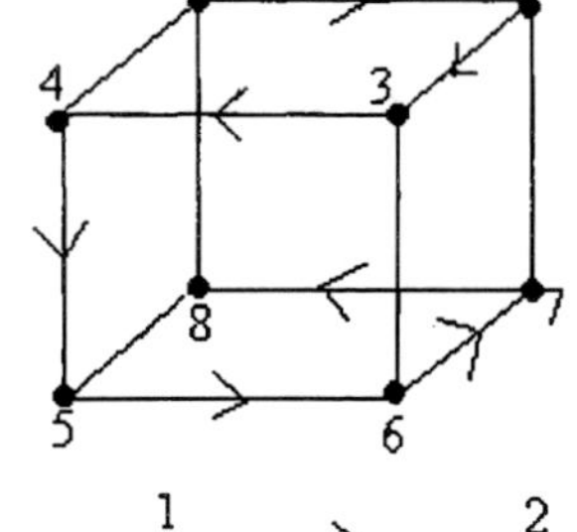

4. This orientation flow shall be taking us through seven steps connecting all the eight corner in a sequence (1, 2, 3, 4, 5, 6, 7, 8, 9).

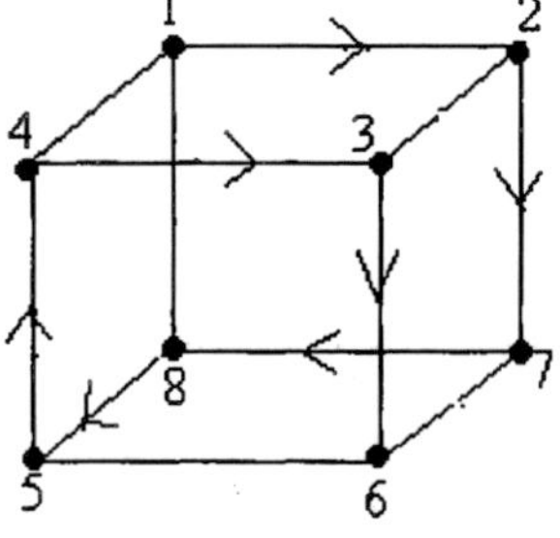

5. We can see that at corner '2', the split for the flow could take along corner '3' or towards corner '7'. The split flow which take from corner '2' towards corner '3' is a flow as has been chased above as (1, 2, 3, 4, 5, 6, 7, 8, 9).

   The split flow which is to take from corner '2' to corner '7' is of chase steps. (1, 2, 7, 8, 5, 4, 3, 6).
6. The flow which has already taken us up till corners 1, 2, 3, 4 and 5, has a split for corner 'eight', as well as for corner '6'. The first flow stream of this split is (1, 2, 3, 4, 5, 6, 7, 8) which has been depicted above in para 4. The second split stream is (1, 2, 3, 4, 5, 8, 7, 6).

(1, 2, 3, 4, 5, 8, 7, 6)

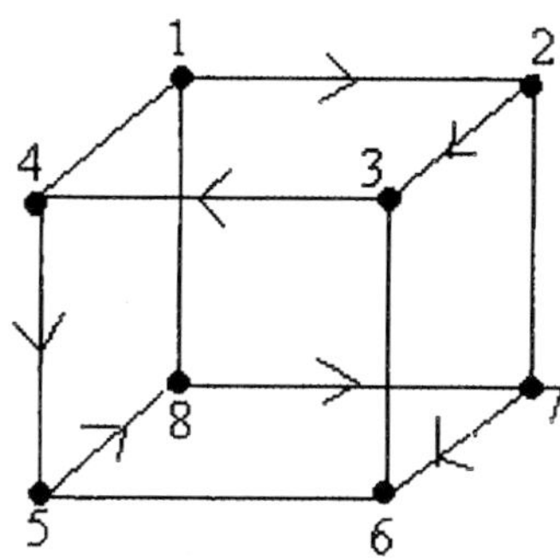

7. The other pair of flow streams would permit depictions as follows

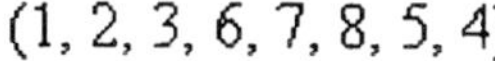

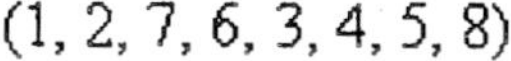

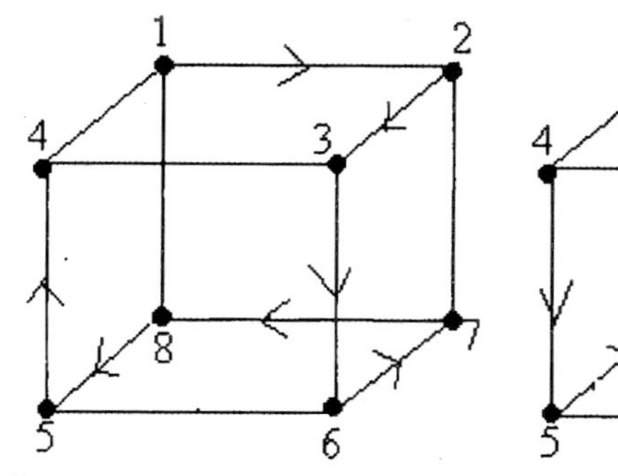

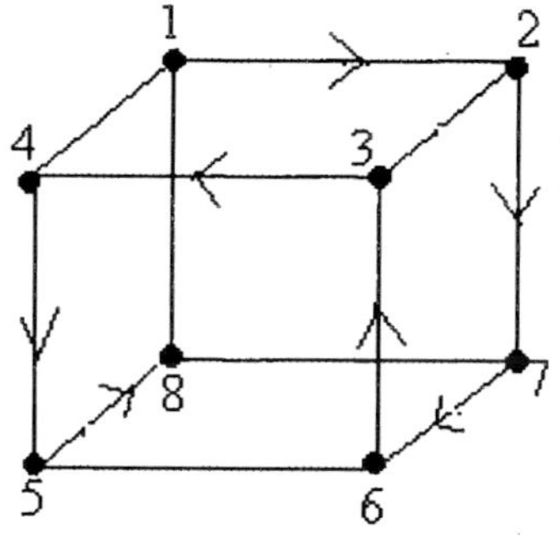

8. This way, it may be comprehended and glimpsed as that the flow starting from corner '1' with orientation along edge leading to corner '2' results into 'five distinct flow streams of seven steps each coordinating eight corner in the following sequences.

Stream	Flow sequence	Flow depiction
1	Corners 1, 2, 3, 4, 5, 6, 7, 8	(1, 2, 3, 4, 5, 6, 7, 8)
2	Corners 1, 2, 7, 8, 5, 4, 3, 6	(1, 2, 7, 8, 5, 4, 3, 6)
3	Corners 1, 2, 3, 4, 5, 8, 7, 6	(1, 2, 3, 4, 5, 8, 7, 6)
4	Corners 1, 2, 3, 6, 7, 8, 5, 4	(1, 2, 3, 6, 7, 8, 5, 4)
5	Corners 1, 2, 7, 6, 3, 4, 5, 8	(1, 2, 7, 6, 3, 4, 5, 8)

9. Now from corner '1' there can be a flow orientation towards '4' or even towards corner '8'.
10. Accordingly, it may be accepted that five flow streams would be available for each of these two orientation as well, and as such, in all there are to follow as many as $5 \times 3 = 15$ flow streams.
11. Then there are eight corners, and as such eight starting point, and hence 15 splits stream from each of the eight corner are to supply us as many as 15 x 8 = 120 splits streams.
12. It would be a blissful exercise to chase all these 120 splits streams.
13. It would be helpful to have a sequence of eight points along the circumference of a circle (1, 2, 3, 4, 5, 6, 7, 8). Being a sequential placement along the circumference it shall be providing a privilege to start from any of these eight placements and to cover the whole range of sequential flow.
14. This shall be yielding '8' settings as follows.

Setting	**Sequential order**
1	(1, 2, 3, 4, 5, 6, 7, 8)
2	(2, 3, 4, 5, 6, 7, 8, 1)
3	(3, 4, 5, 6, 7, 8, 1, 2)
4	(4, 5, 6, 7, 8, 1, 2, 3)
5	(5, 6, 7, 8, 1, 2, 3, 4)
6	(6, 7, 8, 1, 2, 3, 4, 5)
7	(7, 8, 1, 2, 3, 4, 5, 6)
8	(8, 1, 2, 3, 4, 5, 6, 7)

15. It would be interesting exercise to shift from five split flow streams of orientation (corner 1 to corner 2) of setting sequence (1, 2, 3, 4, 5, 6, 7, 8) to parallel five fold split flow streams for other seven settings by simply following the circular sequence order of the settings as follows:

   (i) A shift from setting 1 to setting 2.

Stream	Setting 1	Setting 2
1	(1, 2, 3, 4, 5, 6, 7, 8)	(2, 3, 4, 5, 6, 7, 8, 1)
2	(1, 2, 7, 8, 5, 4, 3, 6)	(2, 3, 8, 1, 6, 5, 4, 7)
3	(1, 2, 3, 4, 5, 8, 7, 6)	(2, 3, 4, 5, 6, 1, 8, 7)
4	(1, 2, 3, 6, 7, 8, 5, 4)	(2, 3, 4, 7, 8, 1, 6, 5)
5	(1, 2, 7, 6, 3, 4, 5, 8)	(2, 3, 8, 7, 4, 5, 6, 1)

Note: All what has been done is that 'one more count' is added at each of the steps. The corners 1, 2, 3, 4, 5, 6, 7, 8 have been changed into 2, 3, 4, 5, 6, 7, 8, 1.

   (ii) Likewise there can be a shift from five fold streams of corner 2 to five fold streams of corner 3. And likewise the process to continue up till a shift from five fold streams of corner seven to five folds streams of corner 8.

(iii) Taking this orientation of corner '1' to corner '2' as orientation of x axis (first axis), there may be a parallel depiction for orientations along y axis and z axis of a three dimensional frame embedded into corner '1'. The students may take the help of the teachers in chase of all these 120 splits stream flow along the edges of the cube.

3.3 BACK TO THE STARTING POINT

1. The seven steps long flow streams running through all the eight corners of the cube, can take back to the starting position but it would amount to re visiting the start with corner.
2. This way the spatial boundary of the cube shall be having as many as 120 circular flow paths and these together shall be making the spatial boundary of the cube as to be of very rich potentialities which deserve to be chased for harnessing its values.

3.4 BACK TO THE SOURCE RESERVOIR

1. The seven steps long flow streams, instead of re-visiting the start with corner, may get connected with the center / origin as the source reservoir.
2. The flow from the source reservoir may reach any of the corners, and its seven steps long flow stream running through all the eight corners shall be taking back to the source reservoir.
3. This way, as many as 120 flow streams originating from the source reservoir / center / origin and

after flowing through all the corner reaching back the source reservoir, is a phenomenon which deserves to be chased.

4. This phenomenon of 120 artifice is structurally very rich set up and is of the order of 120 hyper cubes 4 coverage of boundary of boundary of hyper cube 6.

3.5 TRANSCENDENCE FROM WITHIN

1. This phenomenon of 120 flow streams originating from the source reservoir / center / origin and after running through all the corners and reaching back the source reservoir is a transcendental phenomenon of transcendence from within the center of the cube / origin of 3-space.
2. This transcendental phenomenon of transcendence from within the center of cube / origin of 3-space as seat of hyper cube 4 / 4-space / Creator's space deserves to be comprehended intellectually as well as the same deserves to be experientially glimpsed while permitting the mind to transcend through the spatial boundary of the cube and to reach the center of the cube / origin of 3-space / seat of 4-space and to continue transcending till the new wonderful transcendental worlds are glimpsed.

■

4

CUBE IN MOTION

4.1 TRACK OF A MOVING POINT

1. One must first learn to chase track of a moving point as a single axis phenomenon.
2. It shall be taking to the set up of '1-space as domain' with 0-space in the role of boundary as a pair of end points.
3. This, may formally be defined as the role of 0-space as boundary fold of interval / 1-space.
4. The other distinct roles of 0-space / point is of its own domain, as well as being of the role of dimension of 2-space.
5. The teachers may help the young minds to comprehend different roles of 0-space / points, including it being the origin fold of '2 space'.
6. The parallel existence of dimensional frames and artifices of numbers may be availed for bringing home these fine roles of 0-space and other positive and negative spaces.
7. The conceptual base for –1 space may be brought within comprehension of young minds by

associating with opposite orientation of a line that that in terms of which +1 space is to be depicted.

4.2 TRACK OF A MOVING EDGE

1. The track of a moving edge deserves to be chased in continuity of the track of a moving point and as a sequential step ahead of it.
2. The track of a moving edge, being of the format of the surface plates of a cube, as such the same, by way of definition be accepted and designated as a 2-space set up.

4.3 TRACK OF A MOVING SURFACE PLATE

1. The track of a moving surface plate of the cube deserves to be chased in continuity of the tracks of a moving point and of a moving edge and as a sequential step ahead of them.
2. The track of a moving surface plate, being of the format of the domain of the cube itself, as such the same, by way of definition be accepted and designated as a 3-space set up.

4.4 GRID ZONES: FLOW FROM WITHIN

1. The phenomenon of transcendence from within, in the context of spatial boundary takes us to the center / origin of grid zones as '2-space set ups'.
2. The transcendence through the center / origin of grid zones being seats of 3-space shall be leading to the domain of cube / 3-space.

3 The track of a surface / plane / square / 2-space being a cube / 3-space, as such the grid zones of the surfaces shall be constituting a net of solid grids for the domain of the cube.

4.5 TRACK OF A MOVING SOLID

1. The young minds be helped to chase track of a moving solid / cube as a sequential step in continuity of the moving tracks of point, line and plane.
2. It may be comprehended as that the point as a zero space body, yields a track of 1-space format.
3. Ahead, the line as a 1-space body yields a track of 2-space format, while the plane as a 2-space body yields a track of 3-space format.
4. In continuity, solid / cube as a 3-space body shall be yielding a track of 4-space set up.
5. It may be brought to the conscious notice of the young minds as that they are witnessing the phenomenon of motion of solids with their own motion by walking and jumping and as such the space in which we walk and jump is of a 4-space set up.

4.6 SOLID BOUNDARY OF CENTER/ORIGIN

1. The teachers shall help the young mind to visualize and comprehend as that the center of the cube is enveloped within a solid boundary.
2. Further the teacher shall also the young minds to visualize and comprehend as that, the cube permits cut and split into eight sub cubes, as such the solid boundary of the center consist of eight solid boundary components.
3. The teacher shall further help the young mind to chase cut and split of 3-space as eight octants, parallel to the cut and split of a cube into eight

sub cubes as eight solid components boundary of the center.

4. The teacher shall further help the young minds to comprehend and visualize as that center of the cube is of features parallel to origin of 3-space and as such is a seat of 4-space.
5. The teacher shall further help the young minds to visualize and comprehend center of the cube / origin of 3-space as seat of hyper cube 4 / 4-space.

4.7 SOLID DIMENSIONAL ORDER

1. The teacher shall help the young minds to sequentially take to the role of 3-space as solid dimensional order of 5-space.
2. As a first step, it be focused as that 3 axes of 3-space are of linear format / being lines / 1-space bodies, and as such 1-space is here in the role of dimension of 3-space.
3. The set up of human head equipped with a pair of eyes, may be availed for impressing upon the young minds as that the spatial order of pair of eyes is of the format of 2-space in the role of dimension of 4-space in which we move and jump.
4. As a sequential step, young minds may be helped to visualize and comprehend / 3-space / solids / cube / a head equipped with 3 eyes as a solid format / solid dimensional order of 5-space.

■

5
SOLID QUANTIFIERS

5.1 LINEAR QUANTIFIERS

1. Linear quantifiers by definition be accepted as length units along edges.
2. These are of first power values (A^1).
3. The mathematical tool which helps measure the lengths is designated as a measuring scale.
4. It would be a blissful exercise to construct a measuring scale by having knots at equal distances along a thread.
5. One may have a unit length of ones own choice and in terms of it by having suitable number of knots along a thread, construct a measuring scale of units of ones choice.
6. It would be interesting to have comparisons of measuring scales by comparing the 'unit lengths of the measuring scales.
7. One may acquit oneself with the popular choices of measuring units, illustratively, one centimeter as a unit length or one inch as a unit length, and so on.

8. One may learn to convert one system of units of length, say, 'one centimeter as unit of length' into another system of units of length, say 'one inch as a unit of length'.
9. One may construct the measuring scales and measuring rods of different materials by marking units and fractions of units along them.

5.2 SPATIAL QUANTIFIERS

1. As comparison to linear quantifiers, which accept 'length units', the spatial quantifiers accept 'area units'.
2. While linear quantifiers help measure 'lengths' of lines with the help of 'linear scales' of length units, the spatial quantifiers help measure 'areas' of surfaces with the help of 'linear scales' by enveloping surfaces within linear boundaries, like circumference of a circle.
3. The linear quantifiers are of features of single axis formats as of 'lines'. On the other hand the spatial quantifiers are of features of pair of axis formats of grids frames for its surfaces zooms.
4. The focus of linear quantifiers, in terms of single axis, and of linear units / length units (A^1) is upon the 'domain of interval' / domain of 1-space body / 1-space as domain.
5. The focus of spatial quantifiers, in terms of pair of axes is upon the boundary of square / domain of 2-space body / 2-space as domain.
6. It is this change in the role of 1-space / interval / line / length / linear quantifiers / axis from 'domain' of 1-space / 1-space as domain to its role

as of 1-space set up as boundary of 2-space, which deserves to be chased to reach at the end result of transition from linear quantifiers to spatial quantifiers.

7. Within a set up of a cube, this transition and transformation from linear quantifiers to that of spatial quantifiers attains a shift from the set up of 'edges' to the set up of the 'surfaces plate' framed within 'edges'.
8. This shift is of the features, as that the single axis format results into an orientated flow as of streams, and the remaining at boundary of surface plates and the measuring scales successfully covers the flow as addition of unit counts. On the other hand the shift from boundary (edges) to domain (surface-plates), avails 'pairing of values' along pair of axes by having product thereof and hence the spatial units as to be of second degree powers.
9. The transition from linear quantifiers to spatial quantifiers as such comes to be from Vridhi / addition / first degree power units / A^1 / length units to guna / multiplication / second degree power units / A^2 / area units.

5.3 SOLID QUANTIFIERS

1. The phenomenon of solid quantifiers is of sequential increase as at first step there being a transition from linear to spatial quantifiers being a shift from first degree powers units A^1 to second degree power units A^2, and the same at the second step being from second degree powers units (A^2) to third degree power units (A^3).

2. As the transition from linear quantifiers to spatial quantifiers is attain by shifting from track of moving point to track of a moving line / edge, so in a sequence the transition from spatial quantifiers to solid quantifiers is attained as a shift from the track of a moving edge to the track of a moving surface plates of the cube.
3. The solid quantifiers phenomenon may be chased as transcendence from the centers / origins of grid zones of surface plates constituting boundary of a cube.
4. The solid quantifiers phenomenon may also be chased as solid order at seat of origin of 4-space fulfilling the domain of the cube through its center as origin of 3-space being the seat of Creator's space.
5. It would be a blissful exercise to chase four folds of the set up of a cube with interval / 1-space playing the role of dimension / axis, surface plates / 2-space playing the role of boundary, 3-space content lump playing the role domain of the cube and 4-space playing the role of origin of 3-space / center of a cube.
6. It also would be a blissful exercise to comprehend different roles of 3-space / cube, as its own domain, as boundary of Creator's space / 4-space, and as dimension of transcendental worlds / 5-space.
7. It further would be a blissful exercise to comprehend 3-space / cube in the role of the origin fold of spatial grid zones / 2-space set ups.

5:4 HYPER SOLID QUANTIFIERS

Teachers may help the students to visualize and comprehend the existence and roll of hyper solid quantifiers for chase of the hyper solid domains like hyper cube 4.

5.5 WONDERFUL WORLDS OF REALITY AT REST AT CENTER / ORIGIN.

The end fruit of learning and chase of the set up of the cube would be the bliss of visualizing and comprehending the wonder ful worlds of reality, which other wise are at rest at center of the cube / origin of 3-space being seat of Creator's space / 4-space and whole range of transcendental worlds reality being awaiting the pilgrimage of the transcending minds.

■

For further studies

'Book: Vedic geometry course' may be referred for proper insight into the transcendental phenomenon of emergence and manifestation of the transcendental worlds within creator's space.

■

SECTION - III
GANITA SUTRA SYSTEMS

PART 1

GANITA SUTRAS

3.0 INTRODUCTORY

1. The Ganita Sutras including Upsutras 'supply us first principles for working out the basics.
2. The basics, for the present phase and stage of learning, are supplied by the waking state comprehension of the set up of a cube as representative regular body of 3-space as our world of existence / Triloki / Vishwa / Jagat.
3. The set up of the cube is enveloped within a synthetic boundary of 26 constituents parallel to the features of 26 basic elements which permits chase along artifices of numbers in terms of a pairing of counts as processing operation of chandas / meters / filter measures for coverage of sky between Earth and Sun.
4. Here, the Ganita Sutras system is being taken up for its introductory features in the sequence and order of the Sutras and Upsutras. The focus here is only upon the introduction of the text and its functional rules. The teachers and senior students

who are interested in for in depth chase of the system may refer to the Book of the Author titled 'Vedic mathematics of Ganita Sutras system'.

3.1 GANITA SUTRA-1

3.1.1 TEXT

एकाधिकेन पूर्वेण
Ekadhiken Purvena
One more than before

3.1.2 SEQUENTIAL PROGRESSION RANGE

1	2	3	4	5	6	7	8	9	10
ए	क्	आ	ध्	इ	क्	ए	न्	अ	प्
11	12	13	14	15	16				
ऊ	श्र्	व्	ए	ण्	अ				

3.1.3 FORMAT OUTLINE OF WORKING RULE

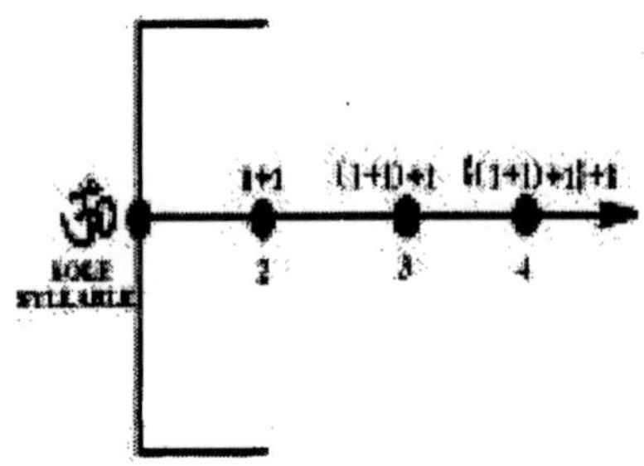

3.1.4 TRANSCENDENTAL PREFIX

Being the first Sutra, it presumes the existence and availability of the transcendental prefix, namely 'Om' being the 'ek akshar Brahman / sole syllable Brahman'.

The Chase of the present phase and stage, is focused upon the vales and features of 'artifices of numbers' and upon parallel geometric formats, as such the value and feature of the transcendental prefix 'Om' being 'ek akshar Brahman / sole syllable Brahman' is focused and is being availed as 'ek / the whole some 'one'.

The value / feature as a start with position 'before', as one, under the functional rule of Ganita Sutra-1 itself, takes from 'one' to 'one more than one', that is, to 'two'.

The wholesome 'one' as artifice '1' is parallel to the set up of 'single axis' of '1-space format', and with it 'one more than one', shall be taking us to 'two', as a pair of axes, and this as such shall be helping us appreciate the soundness and richness of the symbol '+' for plus / addition. This further, simultaneously shall be helping us appreciate the position 'before', as of 'one' as 'one axis', being one less than 'two' and thereby the soundness and richness of symbol '-' for minus.

3.1.5 ORDERING PRINCIPLE

This reach of the functional rule of Ganita Sutra-1which takes us from a single axis to a pair of axis, as such also shall be providing us an insight into the **ordering rule** at the base of the functional rule.

In fact more than the applied value of 'Arithmetic operation of 'addition', the needed focus upon the organizational format and functional rule of Ganita Sutra-1 should be upon the pure and values of ordering principle being supplied by Ganita Sutra-1.

It would be a blissful exercise for the students to enlist as many illustrative situations availing this ordering principle. Beginning with the ordering of counts as sequence of counting numbers / natural numbers / whole numbers / integers and reaching up till the Real numbers line, a step ahead, at complex plane, transcendental solids and super transcendental hyper solids, there is no end in sight of the pure and applied values of this ordering principle.

May it be the computing of age, scaling of quantifiers, tabulation and sequencing of values, equations of different variables, and of different powers, and may it be lines, squares, cubes, hyper cubes, may it be the determinants, matrices, tencers, may it be assigning of roll numbers, framing of time tables, all these and alike there is a very big range and domain of applications of the pure and applied values of the ordering principles of Ganita Sutra-1.

Teachers shall help the young minds to enlist more and more situations accepting the ordering principles for their settings.

3.1.6 TABULATION EXERCISE

The tabulation exercise be undertaken. It may be on the lines:

1. Counting numbers

   1, 2, 3, 4, 5, 6, 7, 8——

2. Powers sequence

   A^1, A^2, A^3, A^4,—

3. Roll numbers

   Roll no. 1, Roll no. 2, Roll no 3,——

4. Variables

   One variable, two variables, three variables and so on. Like that the tabulation to be enriched with addition of 'one by one' entries.

The another way to start, may be

1. Learning one alphabet letter a day.
2. Learning one Sutra a day.
3. Learning one basic feature a day.
4. Learning skills of one basic feature a day.

Like that the tabulation to be enriched with addition of one by one entries of learning.

Still another way to have new set of tabulation may be

1. Knowing about 'trees' one by one.
2. Knowing about 'animals' one by one.
3. Knowing about 'birds' one by one.
4. Knowing about 'elements' one by one.

And like that the tabulation may be enriched. And further like that one by one new starts for such tabulations may be had.

3.1.7 GANITA SUTRA-1 AS SOURCE SUTRA

The organization format of Ganita Sutras (and of Ganita Upsutras) deserves to be chased, as in terms of this chase, one may have the real insight about the functional rules of 'Ganita Sutra system'.

This chase may begin with Ganita Sutra-1 itself, as it being the first step of the system, its organization format, is going to be of features which shall be making it a source Sutra, as much as that, in terms of the sequential steps of the text of Ganita Sutra-1 itself, not only the organization format of Sutra-1 would start unfolding but also that the sequential organization steps of whole range of Ganita Sutras system running from Ganita Sutra-1 to Ganita Sutra –16 (and also of

Ganita Upsutra-1 to Ganita Upsutra-13), as well start unfolding step by step.

It is this feature of the organization format of Ganita Sutra-1 which makes it a 'source sutra'. Here under the salient feature of it, are being introduced but for in depth study, one is to refer to the book of the Author on the topic titled 'Vedic mathematics systems (Ganita Sutras formats).

The chase of the organization formats of Ganita Sutras systems is to begin with the text of Ganita Sutra-1 itself starting with its very first letter.

FEATURES OF FIRST LETTER

1. The first letter of the text of Sutra-1 is 'ए'.
2. 'ए' is the sixth vowel.
3. This, as such, is of features being a 'vowel' and that too being the 'sixth'.
4. Therefore the characteristics of 'vowel' and of artifices 'Six' are attracted and same are to be attended to .
5. Artifice Six is of the order of '6-space'.
6. '6-space' is of the order of Sun/*Divya Pursha*/ *Vishnu*/*Atman*/*Go-Lok*.
7. '6-space' manifests as 'domain' of 'hyper-cube 6'.

FEATURES OF SECOND LETTER

1. The second letter of Sutra-1 is the first *varga* consonant 'क्'.
2. The meanings as per organisation format of this letter (क्) is charmukhi / 4 mouths /Lord Brahma (4-space).
3. *Sathapatya* measuring rod accepts Lord Vishnu (6-

space) as presiding deity of the measuring rod and Lord Brahma (4-space) as presiding deity of the measure of the measuring rod.

4. As such, the first and second letters of the Sutra are interconnected as measuring rod and measure/ domain and dimension.
5. As such the organisation format of *Ganita Sutra*-1, at its first step of first letter (,) is of the features of 6-space as domain while the same at its second step of second letter (क्) is of the features of dimension of 6-space.
6. As such the sequential progression of first two steps takes from domain of 6-space to its dimension being 4-space.

FEATURES OF THIRD LETTER

1. The third letter of *Sutra*-1 is (आ).
2. The organisation format of letter आ = अ + अ.
3. This is known as *Vridhi* / (progression) /elongation /addition / increase.
4. आ = अ + अ is of spatial order (1+1=2).
5. The sequential progression of first two steps takes from 6-space as domain to 4-space as dimension.
6. This sequential progression at its next space takes from dimension (as domain) to dimension of dimension that is from 4-space to its dimension being 2-space.
7. As such the attainment up till this phase and stage is of the order of dimension of dimension, where the dimensional stitching is attainable with transcendental grace.

FEATURES OF FOURTH LETTER

1. The fourth letter of the Sutra is (ध्).
2. However the placements for the fourth and fifth letters are organizationally replaced here as per the completions of compositions rules.
3. The first vowel (अ), as creator, naturally at the phase and stage of the second vowel (इ) is the transcendental Lord.
4. The fourth *Sutra*, accordingly has a working rule (transpose and unite).
5. The letters, 4th and 5th of the text of *Ganita Sutra*-1, are transposed for their placements and then are united as a composition of the text of the Sutra.
6. The letter (ध्) has its placements 4.4 *varga* consonant, that is fourth letter of the fourth row / fourth letter of the fourth column.
7. This placement, as well as the form and frame of the script of this letter, is of a manifestation format of a line within Creator's space, and as such the processing while transforms from transcendence to ascendance within Creator's space (along 4th quarter of *Om* formulation) it ascend beginning with liner sequential order and as such it is of '0' as '1' unit.
8. As such the sequential progression beginning with artifice 6 and being through the phases and stages of artifices 4 and 2 in second and third step, ultimately at its fourth step reach a phase and stage of 2-2=0 and 2^2, 2^1 leading to 2^0 =1.
9. The working rule here being to transpose and apply, as such the transcendental grace synthesises the

transcendental skyline and with it the transcendence automatically transforms into ascendance. Accordingly, the orientation of progression hence forth stands reversed.

FEATURES OF FIFTH LETTER

1. The fifth letter of Sutra is the second vowel (इ).
2. The second vowel as artifice-2 supplies the basic features for this phase and stage of the organisation format of the Sutras.
3. The transposed position of the fifth letter at its fourth letters placed, makes it a letter of double features, as of fourth letter, and hence of artifice 4 /4 space/spatial order and further as of fifth letter and hence of artifice 5 / 5-space.
4. The location of 5-space within 4-space at its origin, and the placement of 5-space within 4-space adds spatial order and hence the split up for solid order of transcendental worlds as northern and southern hemispheres as a pair of 3 dimensional frames of half dimensions.
5. It is this split of three dimensional frame into a pair of three dimensional frames creates a dark zone / zero value place at the origin.
6. It is this placement which brings into '*suniyam*' / zero / '0'.
7. The feature of the zero as a working rule is there as a property of 0 space (dimension of spatial order) which levels all values by the rule $M^0 = 1 = N^0$.
8. The sequential progression which has been unfolding at first step as artifice 6, at second step as artifice 4, at third step as artifice 2 and then at

fourth step as of dual states, being 0 value unit (1) and the reversal of orientation at transcendental skyline, at a next, that is at a fifth step, again attains artifice 2 with availability of the reversal of orientation.

FEATURES OF SIXTH LETTER

1. The letter six of the *Sutra* is the reappearance on the scene of letter (क्) *charmukhi* /Lord Brahma / 4-space /artifice 4, in continuity of the sequential progression order (6, 4, 2, 0 as 2^0, 2).
2. The progression from transcendental skyline from phase and stage of dimension of dimension to the dimension, as such is of internal folds, known as base of un-manifest and un-manifest.
3. As such in proportionate symmetric order, the features of *suniyan* (0) as have been at the phase and stage of fifth step, continue to be their at the phase and stage of the sixth paper as well.

FEATURES OF SEVENTH LETTER

1. The letter seven of the Sutra, in a sequential progression order (6, 4, 2, 0 as 2^0, 2, 4) leads to artifice 6 again and hence the availability of the letter (ए) at seventh step as well.
2. The form and frame of the script of the letter (ए) as is evident of a pair of opposite orientations for the approach for fixation of the middle.
3. It is this feature of this form and frame of the script of letter (ए) which emerges as the source of the features of the organisation format of Sutra-7, and also of its working rule of making available simultaneous handling of addition and subtraction.

FEATURES OF EIGHTH LETTER

1. The letter eight of the Sutra is (न्). Its form and frame manifest the format for *Ganita Sutra*-8 'पूरणपूरणाभ्याम्' *Puranapuranabhyan* / By the completion or non-completion.
2. The script form of letter (न्) well indicates as it extending from middle to one half of the interval. The middle is encircled and is tagged with one half of the interval. Impliedly, it makes the other half of the format of half open interval which is a non completion stage as comparison to the half manifesting the form of the letter (न्) as closed interval format, of completion stage.
3. The closed interval when is to split in a pair of parts, it is to be of non completion stage of half closed interval and of a completion stage of closed interval simultaneously happening as is *Ganita Sutra*-8 ('पूरणपूरणाभ्याम्' *Puranapuranabhyan* / By the completion or non-completion).

FEATURES OF NINTH LETTER

1. The letter nine of the Sutra is (अ), the first *pada* of Braham.
2. It is the ultimate attainment state; a *param gati* / supreme progression.
3. It in continuation of the previous phase and stage of sequential progression of the organisation format of Sutra-8 which accepts format of the feature of split as of completion and non completion states.
4. The artifice 9 accepts re organisation as 3 x 3 of spatial format because of which the domain

boundary ratio is to be of formulation order $A^N : 2^N B^{N-1}$.

5. The completion state parallel to closed interval part of the closed interval, is to be of half unit progression, which as such shall be achieving only half of the boundary of the dimensional bodies. It is this sequential progression of attainment through functional format of half boundary which is of the features of (चलनकलनाभ्याम् *Calana-kalanabhyam* /Differentiation-integral Calculus).

6. The spatial order absorbs reversal of orientation as well. As such, the sequential progression steps 1 to 9 would permit reversal of orientation as steps 9 to 1 and the artifices of the letters 1 to 9 of *Ganita Sutra*-1 would in their reversal shall be yielding text and meanings as under

1	2	3	4	5	6	7	8	9
ए	क्	आ	ध्	इ	क्	ए	न्	अ
On reversal of orientation								
अ	न	ए	क्	इ	ध्	आ	क्	ए

3.1.8 REVERSE ORIENTATION OF TEXT

The reverse orientation text comes to be of sub formulations:

1. अन् 2. एक् 3. इध् 4. आ 5. के

The formulations when chased would lead to the inner folds of the organisation format features of the first half of the formulation.

The first above sub formulation (vu~) well indicates that it is reverse processing for the fourth row of *varga* consonants. The complete sequential progression along it would be (अन्त) which means up till end.

क ख ग ध ङ

च छ ज झ ञ

ट ठ ड ढ़ ण

त थ द ध न

प फ ब भ म

In the context it would be relevant that if this progression would be have been initiated from the first row then it would have been (अंङ्क), which means number.

7. The second formulation of the *Sutra* is *Purvena*. It is a composition of seven letters as under:

1	2	3	4	5	6	7
प्	ऊ	र्	व्	ए	ण्	अ

8. The reverse orientation for the above would be as under yielding sub formulations as follows:

1	2	3	4	5	6	7
प्	ऊ	र्	व्	ए	ण्	अ
1	2	3	4	5	6	7
अ्	ण्	ए	व्	र	ऊ	प्

1. अण् 2. एव् 3. रऊप्

(i) First formulation is first *Mahesvara Sutra*.

(ii) The second formulation means 'also' and third formulations means 'form'.

(iii) Here is the bigger result, as that, the forms (of letters, as well as of formulations, formats and artifices, manifestations and all that) are settled on the format of first *Mahesvara Sutra*.

(iv) The second *Maheshvara Sutra* and the following *Maheshvara Sutras* as well shall be manifesting their forms and frames on the format of first *Maheshvara Sutra* itself.

(v) It is this coding of the basic results as reverse progressions of the Sutras / mantras text which has preserved them as along with their keys.

(vi) This key has given us the leads but for which the chase for the forms and frames of *Devnagri* alphabet letters as well as of the *Maheshvara Sutras* themselves would have, perhaps, been not to be within catch so early and so easily.

(vii) Like that the whole range of *Ganita sutras* formulations is a big resource of basic structural keys preserved there in, in there reverse progression texts.

3.1.9 FULL RANGE OF *GANITA SUTRAS* AND *UPSUTRAS* IS MANIFESTED AS *GANITA SUTRA*-1

Part-1 Sixteen sequential stages of *ganita sutras*

1	2	3	4	5	6	7	8	9	10	11	12	13	14	15	16
ए	क्	टा	ध्	इ	क्	ए	न्	अ	प्	ऊ	श्र्	व्	ए	ण्	ट

1. The sixteen letters of the text of *Ganita Sutra-1* manifest the sixteen sequential stages of the *Sutra*.
2. These sequential stages are the sequential folds of the *Ganita sutras* (including *Ganita Upsutras*).
3. The *Ganita sutras* range of artifice sixteen while *Ganita Upsutras* range is of artifice thirteen along the three points fixation of the range format coordinating, beginning, middle and end.

4. The first sequential stage of manifestation is *Ganita Sutra-1* itself.
5. The sequential stage is expressible as the artifice format of very first letter of the text of *Ganita Sutra-1*, that is, the sixth vowel (ए) .
6. This vowel repeats thrice in the text as first, seventh and fourteenth letter.
7. As such the three specific joints, as beginning, middle and end of the range format emerged to be the first, seventh and fourteenth sutras placements.
8. With it the sequential unfoldment phases and stages of *Ganita Upsutras* 1 to 13 are to run parallel to the artifices of letters 2nd, 3rd, 4th, 5th, 6th, 8th, 9th, 10th, 11th, 12th, 13th, 15th, and 16th respectively.
9. Illustratively, the phase and stage of unfoldment of *Ganita Upsutra-1* is to avail the artifice of second letter of the text of *Ganita Sutra-1*.
10. The *sadhkas* fulfilled with an intensity of urge to know and to chase the unfoldment of the text of *Ganita sutras* (including *Ganita Upsutras*) shall approach the Vedic mathematics systems the Vedic way.
11. The recitation of the twelve transcendental names of *Lord Ganesh* shall be causing initiation of emergence of the transcendental phenomena for the chase of the transcending mind for perfection of intelligence parallel to the attainments of the consciousness states by the *sadhkas*.
12. *Suriya namaskar* shall be connecting the processing systems of the transcending mind with the *Divya*

pursha lively within orb of the Sun with it the transcendence of the mind and its glimpsing of the inner folds and the inner most fold of the transcendental worlds shall be attaining parallel consciousness state supplying quantifiers for the states of intelligence field.

13. The rest is just the play of automation process of God state of consciousness attaining unity state of consciousness and consequential perfection of intelligence.
14. It is this play of automation process which gets initiated with the *sadhkas* availing artifice of sixth vowel during transcendence and attaining perfection of intelligence as well as the sequential unfoldment process of the Brahman order.
15. As such the *sadhkas* fulfilled with the intensity of urge to know and chase along the Vedic mathematics systems shall approach Vedic knowledge the Vedic way and go in 'TRANS' time and again and glimpse the inner folds and inner most fold of the transcendental worlds as a seat of self referral domain of *Vishnu lok/atman/Go-lok/Divya pursha*/God state of consciousness/Format of idol of Lord Vishnu/ Hypercube-6/6-space/ artifice 6/Sixth vowel (ऋ)/*Sathapatya* measuring rod with Lord Brahma, Creator the supreme as presiding deity of the measure.
16. With this the source reservoir of Brahman domain manifesting as Creator's space becomes available for the sadhkas as sole syllable *'Om'* as the source formulation of designations *'Ek akshar Braham'* of four folds parallel to four quarter of Brahman, four

quarter of atman, four Vedas, past, present, future and that transcends them as synthesis of three letters of *Om as Aum*, viz., A, U, M together synthesising a distinct fourth quarter 'AUM'.

17. It is this way *Om* (ॐ) is the source formulation for chase of *Ganita Sutra-1*.

18. As such the chase of manifestation of *Ganita Sutra-1* along its transcendental base, is to begin with tapping of the source reservoir along the artifices of source formulation *Om* (ॐ) by approaching it quarter by quarter and the fourth quarter as synthesis of the first three quarter.

19. This chase of four quarter manifestation beginning with our *Triloki* (3-space) takes us uptil *Suriya* (6-space).

20. The first three quarter together manifest the range (3-space, 4-space, 5-space) and the last fourth quarter manifesting as (6-space).

21. The linear order as dimension of dimension of transcendental worlds (5-space) transforms as spatial order of dimension of dimension of self referral domain (6-space).

22. The different roles of 6-space as manifestation layers folds may be tabulated as under:

Manifestation Layer	Dimension Fold	Boundary Fold	Domain Fold	Origin Fold
5th	3-space	4-space	5-space	6-space
6th	4-space	5-space	6-space	7-space
7th	5-space	6-space	7-space	8-space
8th	6-space	7-space	8-space	9-space

23. This background of different features of Vishnu *lok/atman/Divya pursha/Go-Lok*/Format of idol of Lord Vishnu/Hyper cube-6/6-space/artifice 6/ *Pursha format/Sathapatya* measuring rod/6th vowel may help the sadhkas to conveyance themselves as to the need of intellectual chase as well as the need of availing this artifice format of 6th vowel for transcendence and glimpsing of the inner folds and inner most fold of the transcendental worlds, for glimpsing the seat of self referral domain within the transcendental worlds, for fulfilling the mind with ambrosia of bliss of the God State of consciousness lively at the seat of self referral domain within transcendental worlds as there ultimate fold.

24. The intellectual chase of artifice-6 shall be focusing upon the features of this artifice accepting organisation as $1+2+3=1\times2\times3=2+2+2=3+3=6$ which to permit chase along monad (as 6), along di-monad (as 3+3), along tri-monad (as 2+2+2) and further the sequential increase feature as merger of addition and multiplication operation (being $1+2+3=1\times2\times3$).

25. These are just initial stage features of this artifice. The further chase of this artifice by transcendence to the format beneath this artifice as of the order of 6-space shall be sequentially leading to features of artifices 06, 16, 26, 36, 46, 56, 66, 76 and so on.

26. These artifices with 6 at unit place are manifesting distinct features of 6-space as much as that $16=5+6+5$ is linear expression for 6-space as domain within 5-space as boundary.

27. This feature, as such shall be helping the chase of the sequential phases and stages of organisation format of *Ganita Sutra-1* on the one hand, and the whole range of 16 *Ganita sutras* on the other hand.
28. Here in the context it would be relevant to mention as that the artifice 13 of the range of 13 *Ganita sutras* admits accepts linear expression for 5-space as domain within 4-space as boundary (13=4+5+4).
29. This as such shall be jointly manifesting 16 *Ganita sutras* and 13 *Ganita Upsutras* as Hypercube-6 within transcendental boundary manifesting as hypercube-5.
30. The artifice 26 is parallel to 26 basic elements of Vishnu lok. And, like that, sequentially the chase shall be unfolding different phases and stages of unfoldment of different features.

Part-2 Thirteen sequential stages of upsutras

1. The sixth vowel (ए) marks its presence thrice as first, seventh and fourteenth letter of the text of *Ganita Sutra-1*.
2. First Sutra is of working rule (one more than before) while the fourteenth Sutra is of working rule (one less than before).
3. These rules supply the formats for addition and subtraction operation respectively. The seventh Sutra (***Sankalana-Vyaykalna-Bhayam***) is simultaneous addition and subtraction.
4. These three placements stitch the range at these three places as beginning, middle and end. Along this format stand organize the thirteen *Gania Upsutras* as thirteen sequential steps.

5. The artifices sixteen and thirteen respectively accept organisation arrangements with coordination of artifices as 5+6+5 and 4+5+4.
6. Parallel to these manifest hypercube-6 domain within 5-space boundary and hypercube-5 domain within 4-space boundary. These together manifest hypercube-6 within hypercube-5.
7. The artifice 13 as 13 edged cube is hypercube-4 manifesting within Creator's space.
8. The spatial order of Creator's space as a pairing operation works out 13+13=26 basic elements formats of *Vishnu lok* (6-space) with spatial order of Creator's space being dimension of dimension of 6-space.
9. This way *Ganita sutras* take the features of sixteen *Ganita Sutra* a step ahead as 26 basic elements order of 6-space attainable through pairing operation within Creator's space.
10. With it stands established the *Sathapatya* measuring rod with Lord Vishnu as its presiding deity provide Lord Brahma, Creator the supreme being the presiding deity of the measure of this measuring rod.
11. The chase of 16 sequential steps of the range of 16 *Ganita sutras* and the chase of 13 sequential steps of the range of 13 *Ganita Upsutras* may be chased independently as well as simultaneously along the format of hypercube-6 as representative regular body of 6-space.
12. The *Ganita sutras* range of 16 sequential steps may be chased with focus upon 6-space in the role of

domain but manifesting within transcendental boundary (5-space).

13. The *Ganita Upsutras* range of 13 sequential steps may be chased with focus upon 5-space in the role of boundary but within a manifested boundary (4-space).
14. The three point fixation of the base in terms of artifices of 6th vowel as first, seventh and fourteenth letter respectively, as beginning, middle and end points puts forward three different ways for chase of the sequential steps for both *Ganita sutras* as well as *Ganita Upsutras*.
15. The first way is as is being taken up above of beginning with the first letter and reaching uptil the sixteenth letter as the last letter.
16. The second way is to begin from the middle and proceed towards either end. The third way is to begin with the last letter and reach uptil the first letter.
17. This way, it may be evident, as that the organisation format of *Ganita sutras* and *Upsutras* is structurally very rich. The full features of this organisation format can only be glimpsed by the transcending mind.
18. Once these features are to be experientially glimpsed and it is only thereafter that those can be intellectually chased in terms of perfected intelligence.
19. As such the sadhkas fulfilled with an intensity of urge to know and to chase the full features of organisational format of *Ganita sutras* shall permit the mind to transcend availing the artifices of

Ganita sutras and *Upsutras* and to glimpse the inner folds and inner most fold of the transcendental worlds and to attain perfection of intelligence with the privilege of automation process of the God state of consciousness of self referral domain lively as the ultimate fold of the transcendental worlds.

20. Initially the sadhkas shall transcend along the artifices of the text of *Ganita Sutra*-1 and thereafter to sequentially continue chase along the artifices of the other texts of Sutras and *Upsutras*. However here under this chase is being continued in respect of Ganita Upsutras while the chase for Ganita Sutras is to be taken at next phase and stage of study at high school mathematics.

■

PART-II
GANITA UPSUTRAS

1 GANITA UPSUTRA-1

1.1 TEXT

प्रणवः । आनुरूप्येण

PARNAVA ANURUPYENA

1.2 SEQUENTIAL PROGRESSION STEP: '1'

1	2	3	4	5	6	7	8	9	10
आ	न्	उ	श्र	ऊ	प्	;~	,	ण्	अ

1.3 FEATURES SOURCES

1. '*Om*' as *Upsutras* are part of the *Ganita Sutras*.
2. '*Pranava*' as *Upsutras* also constitute a self contained scripture.
3. Artifice 2 as *Upsutras* being second part, as the first being Sutras.
4. Artifice 1 as it is first Upsutra and as such first sequential progression step.
5. *Ganita Sutra*-1 as, to it, it follows as a sub Sutra.
6. Letter 'क्' as the second letter of *Ganita Sutra*-1.

7. Comprehension approach chase of this phase and stage.
   (a) Parents attention along *Pursha format*
   (b) Young minds focus along artifices coordination
   (c) Sadhkas initiatives for transcendence
   (d) Teachers help with artifices formats
   (e) Seniors sharing of GLIMPSING experiences

1.4 TECHNICAL TERMS MEANINGS

1. The formulation आनुरूप्येण *Anurupyena* accepts a pair of sub-formulations (i) आनु *Anu* and (ii) रूप्येण *Rupyena.*
2. The formulation आनु *Anu* means 'to follow'. The formulation रूप्येण *Rupyena* means 'as the form as it is within ण् (nh) *anubandha*/framed.
3. The working rule comes to be 'to follow the form as it is framed'.

2. GANITA UPSUTRA-2

2.1 TEXT

शिष्यते शेषसंज्ञः

Sisyate Sesasamjnah

2.2 SEQUENTIAL PROGRESSION STEP: '2'

1	2	3	4	5	6	7	8	9	10
श्	इ	ष्	य्	अ	त्	ए	श्	ए	ष्
11	12	13	14	15	16	17	18		
अ	स्	अ	ं	ज्	ञ्	अ	:		

2.3 FEATURES SOURCES

1. '*Om*' as *Upsutras* are part of the *Ganita Sutras*.
2. '*Pranava*' as *Upsutras* also constitute a self contained scripture
3. Artifice 2 as *Upsutras* being second part, as the first being Sutras.
4. *Ganita Upsutra*-1 as, to it, follows Upsutra-2.
5. Letter 'आ' as a third letter of *Ganita Sutra*-1.
6. Comprehension approach chase of this phase and stage.
7. Parents attention along *Pursha format*
8. Young minds focus along artifices coordination
9. Sadhkas initiatives for transcendence
10. Teachers help with artifices formats
11. Seniors sharing of GLIMPSING experiences

2.4 TECHNICAL TERMS MEANINGS

The formulation 'षिश्यते षेशसंज्ञः / *Sisyate Sesasamjnah*' avails sub formulations (i) षिश्यते and (ii) षेशसंज्ञः. Meanings rendering for these formulations comes to be 'remainder षिश्यते (i) is designated संज्ञः (transcendental) as that remains षेश (after operation) at the base (ii)'. Conceptually, it means (that), 'that which is to remain un manifest, that the same goes to the transcendental base of manifestation within frames'.

3 GANITA UPSUTRA-3

3.1 TEXT

आद्यमाद्येनान्त्यमन्त्येन

ADYAMADYENATYAMANTYENA

3.2 SEQUENTIAL PROGRESSION STEP: '3'

1	2	3	4	5	6	7	8	9	10
आ	द्	य्	अ	म्	आ	द्	य्	ए	न्
11	12	13	14	15	16	17	18	19	20
आ	न्	त्	य्	अ	म्	अ	न्	त्	य्
21	22	23							
ए	न्	अ							

3.3 FEATURES SOURCES

1. '*Om*' as *Upsutras* are part of the *Ganita Sutras*.
2. '*Pranava*' as *Upsutras* also constitute a self contained scripture
3. Artifice 2 as *Upsutras* being second part, as the first being Sutras.
4. Artifice 3 as it is third upsutra and as such third progression step.
5. *Ganita Upsutra*-2 as, to it, follows Upsutra-3.
6. Letter 'इ' as a fourth letter of *Ganita Sutra*-1.
7. Comprehension approach chase of this phase and stage.
8. Parents attention along *Pursha format*
9. Young minds focus along artifices coordination
10. Sadhkas initiatives for transcendence
11. Teachers help with artifices formats
12. Seniors sharing of GLIMPSING experiences

3.4 TECHNICAL TERM MEANINGS

1. The 'आद्यमाद्येनान्त्यमन्त्येन' = 'आद्यमाद्येन्अन्त्यमन्त्येन' is artifice formulation of two parts, namely, (i) 'आद्यमाद्येन' and (ii) 'अन्त्यमन्त्येन'.
2. Both these parts again are of two sub parts each namely (i) 'आद्यमाद्येन' = (ia) 'आद्यम्' + (ib) 'आद्येन' and (ii) 'अन्त्यमन्त्येन' = (iia) 'अन्त्यम्' + (iib) 'अन्त्येन्'.
3. It may be focused here as that the first part begins with 'आ' and also accepts its both sub parts as well beginning with 'आ'. While on the other hand the second part begins with 'अ' and its both sub parts as well begin with 'अ'. The combination rules of vowels are, as that (A) अ + अ = आ and (B) आ + आ = आ.
4. The simple English rendering for the first part as of a pair of sub parts (ia) 'आद्यम्' + (ib) 'आद्येन' would be 'first' 'with the first'. Likewise the simple English rendering for the second part as a pair of sub parts (iia) 'अन्त्यम्' + (iib) 'अन्त्येन्' would be 'last' 'with the last'.

4 GANITA UPSUTRA-4

4.1 TEXT

केवलैः सप्तकं गुण्यात् || 4 ||

KAIVALAEH SAPTAKAM GUNYAAT

4.2 SEQUENTIAL PROGRESSION STEP: '4'

1	2	3	4	5	6	7	8	9	10
क्	ए	व्	अ	ल्	ऐ	:	स्	अ्	प्
11	12	13	14	15	16	17	18	19	20
त्	अ	क्	अ	ं	ग्	अ	ण्	य्	आ
21									
त्									

4.3 FEATURES SOURCES

1. '*Om*' as *Upsutras* are part of the *Ganita Sutras*.
2. '*Pranava*' as *Upsutras* also constitute a self contained scripture
3. Artifice 2 as *Upsutras* being second part, as the first being Sutras.
4. Artifice 4 as it is fourth *Upsutra* and as such third progression step.
5. *Ganita Upsutra*-3 as, to it, follows Upsutra-4.
6. Letter '/k' as a fifth letter of *Ganita Sutra*-1.
7. Comprehension approach chase of this phase and stage.
8. Parents attention along *Pursha format*
9. Young minds focus along artifices coordination
10. Sadhkas initiatives for transcendence
11. Teachers help with artifices formats
12. Seniors sharing of GLIMPSING experiences

4.4 TECHNICAL TERMS MEANINGS

1. The terms 'केवलैः सप्तकं गुण्यात् *Kevalaih Saptakam Gunyat*', in its three sub formulation (i) केवलैः (ii) सप्तकं (iii) गुण्यात् ,straight a way accepts simple English rendering for them as (i) only, (ii) seven and (iii) degrees (features/multiplicative/properties) which together to mean as that 'it is only uptil the seventh degree that the features are to lead'.
2. It is to be comprehended as that the artifice 5 is to inherently supply the features and parallel to it space 5 which in the role of dimension to lead uptil seven space.

3. This shift from manifestation format to its transcendental base is to focus upon the *yoga nishtha* processing features. The chase of transcending mind by itself is to be in terms of the quantifiers to be supplied by the consciousness field to the intelligence field and this as such would mean to avail the artifices parallel to the dimensional formats. It is known as coding values for the transcendental features. It is this precisely which is the basic focus of features of this phase and stage of the organisation format of *Ganita Upsutras*.
4. The Vedic code of artifices of numbers for the alphabet letters otherwise availing geometric formats is at work here and the same on its application works out the fractions 1/7 as sequential units '142857'.
5. As parallel to the artifices of numbers manifest dimensional frame and spaces, as such the organisation formats accordingly also acquire the features of dimensional frames as well as of dimensional spaces for the corresponding phases and stages of respective serials of steps, illustratively, the code value for letter 'क्' is '1', for letter 'व्' is '4' and for letter 'ल्' is '3' and the formulation 'केवलैः' amounts to coded value '143'. The feature that the multiplicand for seven being '143' makes the organisation format of *Ganita Upsutra* 'केवलैः सप्तकं गुण्यात' as to be of its wonders of Ancient wisdom compilation techniques.
6. The value for 1/7 as sequential units of range '142857' with 142 and 857 being interconnected

as 1+8=9, 4+5=9 and 2+7=9; and 142+001=143 and 1000=999+001

5 GANITA UPSUTRA-5

5.1 TEXT

वेष्टनम्

VESTANAM

5.2 SEQUENTIAL PROGRESSION STEP: '5'

1	2	3	4	5	6	7	8
व्	ए	ष	ट्	अ	न्	अ	म्

5.3 FEATURES SOURCES

1. '*Om*' as *Upsutras* are part of the *Ganita Sutras*.
2. '*Pranava*' as *Upsutras* also constitute a self contained scripture
3. Artifice 2 as *Upsutras* being second part, as the first being Sutras.
4. Artifice 5 as it is fifth *Upsutra* and as such third progression step.
5. *Ganita Upsutra*-4 as, to it, follows Upsutra-5.
6. Letter 'क्' as a sixth letter of *Ganita Sutra*-1.
7. Comprehension approach chase of this phase and stage.
8. Parents attention along *Pursha format*
9. Young minds focus along artifices coordination
10. Sadhkas initiatives for transcendence
11. Teachers help with artifices formats
12. Seniors sharing of GLIMPSING experiences

5.4 TECHNICAL TERM MEANINGS

1. The वेश्टनम् *Vestanam* is a technical term. It means so positioned (व) as air within space to continuously salute (नम) the transcendental presiding deity (इश्ट); the transcendental presiding deity (इ . श्ट), the double seat at origin of 4-space (अ), as 5-space (इ) and as 6-space (श्ट of artifice 6); 4=2+2=2×2=-2×-2 and 5=2+3, and 6=2×3=-2×-3.
2. It is osculation, a process of first half synthesising with second half but crosswise as first with the first and last with the last to avoid tearing of at the base and to avail continuity straight of first half with the second half already attained uptil the previous sequential step (4) of the organisation format of *Ganita Upsutras*.

Transition from Sequential step 4

3. The sequential step (4) along the artifices of numbers for the value 1/7 as unit range 142857 with first half 142 and second half 857 as complementary and supplementary of first so synthesised as the full range 142857 and having cross coordination for first of first range with first of the second range (1 & 8) as well as for last of first and last of second (2 & 7) and also emerging for the middle of both (4 & 5) with affine values range 9,9,9 as the place value left behind after lifting of the digits for placements along the place value.
4. It is this attainment of this phase and stage which makes osculators as the cosmic features of the transcendental order unfolding wonders of artifices along ten place value system.

5. It would be a blissful exercise for the sadhkas to transcend time and again and experience the bliss of cosmic features of transcendental order being put in place by the osculators for the decimal expressions for the fractions.
6. One may be proud of ones attainments in terms of the range of osculators being chased by him. As long would be this range, so high would be is linear progression attainments along the *Ganita Sutras* and *Upsutras* organisation format. For this the sadhkas shall begin with the osculator chase for 1/7=7/49; 4+1 as *Ekadhika* and hence 5 as osculator.
7. For perfection of comprehension of the transition from organisation format of specific osculator feature of fourth step to the general osculators feature of the fifth step, it would be a blissful exercise for the sadhkas to transcend time and again and to chase organisation format of this phase and stage of fifth step beginning with the organisation format of fourth step as:
   (i) The orientation flow from corner 1 to the last (8th) coordinated through edges 1 to 7 works out reversal of orientation of top surface (framed within first 3-edges) while reaching at the bottom surface (framed within the last 3-edges namely 5th, 6th and 7th edges).
   (ii) It is this coordination of pair of surfaces of opposite orientation which is of special feature of organisation format of fourth sutra.
   (iii) This special feature deserve to be comprehended with pointed focus, firstly as

to the reversal of orientation of top and bottom surfaces, secondly as that the middle edge (4th edge) permits reversal of orientation without disturbing the organisation format but attaining reversal from bottom to top through the fourth edge itself. It is this feature, which take to the next phase and stage of organisation format of *Ganita Upsutras*, namely, to the phase and stage of fifth *Ganita Upsutra* (वेश्टनम् *VAISHETANAN*) which means osculators/cyclic orders.

(iv) The transition from artifice format of *Ganita Upsutra*-4 to *Ganita Upsutra*-5 deserves to be chase as gently as it deserve, a transition from spatial order of pair of parts to tri-moand format, at the base of tri-monad format whose joint is just at '0' value as sustained by un manifest base as third (central) part of tri monad.

(v) It is this coordination of reversal of orientation through '0' value joint which deserves to be comprehended. It is parallel to the format for synthesis of left and right parts of human frame.

(vi) *Ganita Upsutra*-4 works out value of 1/7 as recurring six digits range '142857' of which the first three digits sub range '142' is of compliments from 9's of second part sub range namely '857' permitting chase as '9-8=1', '9-5=4' and '9-7=2'. The zero value joint synthesises both parts as full range '142857' as a unit range, and unit after unit extending the complete infinite range of full value expression for 1/7.

(vii) These parts, as a pair of parts of a circle shall be coordinating the respective digits of both sub ranges crosswise, as much as that 1 & 8 would be coordinated and like wise 4 & 5, as well as, 2 & 7 shall be coordinated.

(viii) The positions allocations, once fixed for 6 digits as (1) 1, (2) 4, (3) 2, (4) 8, (5) 5 and (6) 7, then the order of values of these six digits 1, 4, 2, 8, 5, 7 to be sequentially arranged as (1)1, (2) 2, (3)4, (4) 5, (5)7 and (6) 8. These two orders are to be simultaneously chased for values of 1/7, 2/7, 3/7, 4/7, 5/7 and 6/7 and the same shall be yielding as

1/7=(142857)

2/7=(285714)

3/7=(428571)

4/7=(571428)

5/7=(714285)

6/7=(857142)

6. GANITA UPSUTRA-6

6.1 TEXT

यावदूनं तावदूनम्

Yavadunam Tavadunam

6.2 SEQUENTIAL PROGRESSION STEP: '6'

1	2	3	4	5	6	7	8	9	10
य्	आ	व्	अ	द्	ऊ	न्	अ	म्	त्
11	12	13	14	15	16	17	18		
आ	व्	अ	द्	ऊ	न्	अ	म्		

6.3 SOURCE FORMULATION *OM*

1. The source formulation *Om* is of four *padas* parallel to four *padas* of atman/self, four *padas* of *Brahman*, four Vedas and four mouths of Lord Brahma, Creator the supreme.
2. The sequential progression through the four *padas* of *Om* formulation, in its forth phase and stage, attains transition and transformation from the combined set up of first three *padas* . It is this transition and transformation, which makes Creator's space, a spatial order space.
3. As such this attainment of reaching Creator's space requires re processing all through to be of the features of the spatial order. It is this processing which had begun with the artifices format of *Ganita Upsutra*-5.
4. The hyper cube 4, a representative regular body of 4 space, accepts five non negative and geometries. It is this attainment, which takes coverage of five steps and accordingly the reach up till step 5 had been there. Simultaneously the spatial order of Creator's space initiates processing again at the fifth step.
5. This amounts to transition from the manifestation layer (1, 2, 3, 4) to the next manifestation layer

(2, 3, 4, 5). With this the processing reach in its fourth quarter is to the transcendental worlds (5-space) of solid order but within four space and as such the transcendence takes place in two steps, that is, from domain to dimension and then from dimension to dimension of dimension.

6. It is this double transcendence feature which is at the base of the organisation format of *Ganita Upsutra*-6.

6.4 TECHNICAL TERMS AND MEANINGS

1. The Upsutra text 'यावदूनं तावदूनम् *Yavadunam Tavadunam*' avails a pair of terms '(i) यावदून' and '(ii) तावदूनम्ष' The chase of these terms takes to sub terms for first term being 1(a) 'यावत' and 1(b) 'उनम्' and to sub terms for the second term being 2(a) 'तावत' and 2(b) 'उनम्'.
2. The chase of sub terms of first term leads to 1(a) 'यावत'= 'या' + 'वत्' and 1(b) 'उनम्'. The simple English rendering for these formulations is 'from this, the deficiency'. Likewise the chase for the second term would lead to 2 (a) 'तावत' = 'ता' + 'वत्', and 2(b) 'उनम्' with simple English rendering for them as 'from that, the deficiency.'
3. The combined working rule comes to be, 'from this (given value) the deficiency (from the base value) is to be worked further for that obtained value the further deficiency'. This, illustratively for 8 for base 10, is to as a first step is reaching at 10-2 and the second step is reaching at (10-2)-2. Along the geometric formats for the transcendence rule, it means, as a first step reaching from domain to dimension and then as a next step reaching from dimension of dimension.

4. Within transcendental worlds its amount to reaching from 5 space to 3 space and further from 3 space to 1 space. This in Vedic language means transcendental chase of *Lord Ganesha* firstly with his all the five heads and then with three heads and one head respectively to glimpse 'the sky line of the transcendental worlds constituted by dimensions of dimensions' to attain transcendental grace for transcending the sky line of transcendental worlds to attain self referral feature of the self referral domains of *Vishnu lok* (6 space).

7. GANITA UPSUTRA-7

7.1 TEXT

यावदूनं तावदूनीकष्त्य वर्गं च योजयेत्

Yavadunam Tavadunikrtya Varganca Yojayet

7.2 SEQUENTIAL PROGRESSION STEP: '7'

1	2	3	4	5	6	7	8	9	10
य्	आ	व्	अ	द्	ऊ	न्	अ	ण	त्
11	12	13	14	15	16	17	18	19	20
आ	व्	अ	द्	ऊ	न्	इ	क्	ऋ	त्
21	22	23	24	25	26	27	28	29	30
यृ	अ	व्	अ	र	ग्	अ	ण	च्	अ
31	32	33	34	35	36	37			
य्	ओ	ज्	अ	य्	ए	त्			

7.5 TECHNICAL TERMS MEANINGS

1. The text of the *Upsutras*-7 avails the set of five technical terms/formulations, namely, (i) यावदून (ii) तावदूनीकष्त्य (iii) वर्ग (iv) च and (v) योजयेत्.
2. Of these, the basic terms यावदूनं and तावदूनम् have already being chased in previous sequential progression step. The above second technical term rkonwuhd'R; avails a pair of sub formulations, of which, the first sub formulation is of the features 'तावदूनम्' while the second sub formulation, namely, 'कष्त्य' accepts simple rendering of 'reaching at, as creative step' which means to have reached at the deficiency in it's a pair of steps, it is to be availed, in two fold manner, firstly as such, and secondly as second degree for the deficiency.
3. It is of the feature of 'double' and of 'square'. The third term वर्ग means square/squaring. It this way makes specific of the use for 'squaring as well' obtained by च योजयेत् (addition/also/as well).
4. The combined rule is of reaching at a SQUARE by availing the two fold feature of **'deficiency from the base'** as being 'double' and also being 'square' applications. The often cited illustration is of reaching at square of eight by availing its deficiency from base as 'two', which of its square feature is 'four' for first digit and of its double deficiency from the base is of feature 'ten minus two minus two' as six for the second digit and thereby being the value for the square of eight as sixty four.

8. GANITA UPSUTRA-8

8.1 TEXT

अन्त्ययोर्दषकेष्पि

Antyayordasake 'pi'

8.2 SEQUENTIAL PROGRESSION STEP: '8'

1	2	3	4	5	6	7	8	9	10
अ	न्	त्	य्	अ	य्	ओ	द्	अ	र्
11	12	13	14	15	16	17			
श्	अ	क्	ए	(ह)	प्	ठ			

8.3 TECHNICAL TERMS MEANINGS

1. The term 'अन्त्ययोर्दषकेष्पि' is a composite term of sub formulation which permits chase as 'अन्त्ययोर्दषकेष्पि' = अन्त्य् + अय् + ओर् + एव + दष + के + अपि which on its chase would mean **'the (अन्त्य) end on its reverse (अय्) orientation (ओर्) leads too (एव) ten *chatarmukhi / Brahmas* (d) as well.** This means as that the transcendental grace at the sky line embeds spatial order for the sky line manifesting as spatial order boundary of ten components of the order of Creator's space; a manifestation format of hyper cube 5, the representative regular body of 5 space within 4-space.

9. GANITA UPSUTRA-9

9.1 TEXT

अन्त्ययोरेव

Antyayoreva

9.2 SEQUENTIAL PROGRESSION STEP: '9'

1	2	3	4	5	6	7	8	9	10
अ	न्	त्	य्	अ	य्	ओ	र्	ए	व्
11									
अ									

9.3 TECHNICAL TERM MEANINGS

1. The अन्त्ययोरेव / *Antyayoreva* is a technical term for whose meanings, its composition which admits split-up as अन्त्य् + अय् + ओर् + एव which on its chase would mean **'the (अन्त्य) end on its reverse (अय्) orientation (ओर्) leads too (एव)'.** The focus here is upon the feature of 'end point' of sequential set up, as that there happens reverse of orientation for continuation of further leads (of sequential progressions).
2. This reversal of orientation availability for continuity of sequential progression is the spatial feature of the organisation order of Creator's space (4-space) and when the same is manifesting boundary of transcendental worlds (5-space), it, in a way is to take from boundary to the transcendental worlds themselves.
3. It is this feature of chase which becomes the basic feature of organisation format of this phase and stage of *Ganita Upsutra*-9. From it naturally the further attainment is to be the transition of attainment of equal heights through summation in affine set ups with removal इतः /(*itah*)/ transcendence from गुणः *gunas*/ multiplicative features of summations (additions). It would be a

phase and stage of transcendence from the solid order (third degree order) of the transcendental worlds (5-space with 3-space as dimension) to the hyper worlds (6-space with 4-space as dimension); the phase and stage of features (निस्त्रै गुण:).

10 GANITA UPSUTRA-10

10.1 TEXT

समुच्चयगुणित:

Samuccayagunitah

10.1 SEQUENTIAL PROGRESSION STEP: '10'

1	2	3	4	5	6	7	8	9	10
स्	अ	म्	उ	च्	च्	अ	य्	अ	ग्
11	12	13	14	15	16				
उ	ण्	इ	त्	अ	:				

10.3 TECHNICAL TERMS MEANINGS

It is this functional change of continuing progression with only one of the two parts of di monad format which deserves to be comprehended and chased with conscious focus.

This doubling and halving features handling as $1=2\times1/2$ attains stripping of the multiplicative properties of all degrees and powers.

It is this unique attainment to which the artifice format of *Ganita Upsutra*-10 leads to being 'समुच्चयगुणित: *Samuccayagunitah*', as that fully (leqPp;) the multiplicative attributes (गुण:) stand stripped off (इत:) and with it affine state is attained as of Creator's order with transcendental base/transcendental origin (5 space).

11. GANITA UPSUTRA-11

11.1 TEXT

लोपनस्थापनाभ्याम्

Lopanasthapanabhyam

11.2 SEQUENTIAL PROGRESSION STEP: '11'

1	2	3	4	5	6	7	8	9	10
ल्	ओ	प्	अ	न्	अ	स्	थ्	आ	प्
11	12	13	14	15	16	17			
अ	न्	आ	भ्	य्	आ	म्			

11.3 TECHNICAL TERM MEANINGS

This attainment of affine state for Creator's order (4-space order) with transcendental base (5-space) with functional progressions availing only one part of the pair of parts of spatial order of di monad format at a time would amount to a phenomenon of 'DISAPPEARANCE AND REAPPEARANCE' (लोपनस्थापनाभ्याम् *Lopanasthapanabhyam*) of the transcendental features for the transcending mind till the transcending mind glimpses the inner most fold of the Creator's space as a seat of the transcendental worlds an further glimpses the inner most fold of the transcendental worlds as a seat of self referral domain and thereby there being the attainment of perfection of intelligence through GLIMPSING (विलोकनम् *Vilokanam*).

This phase and stage of लोपनस्थापनाभ्याम् *Lopanasthapanabhyam*/'**DISAPPEARANCE AND**

REAPPEARANCE' of transcendental features for the transcending mind is to be successfully through by constantly reminding one self of the transcendental base of the whole range of manifested creations causing such **'DISAPPEARANCE AND REAPPEARANCE'**.

12. GANITA UPSUTRA-12

12.1 TEXT

विलोकनम्

Vilokanam

12.2 SEQUENTIAL PROGRESSION STEP: '12'

1	2	3	4	5	6	7	8	9
व्	इ	ल्	ओ	क्	अ	न्	अ	म्

12.3 TECHNICAL TERMS MEANINGS

1. The term विलोकनम्/Vilokanam accepts split as fo + लोक + नम्. And that this is to be transcended for **GLIMPSING** of transcendental (वि) worlds (लोक) for their salutations (नम्) and as a reward there of to have perfection of intelligence of the order which chases गुणितः = गुणः + इतः/stripping off of all multiplicative feature as full affine state (समुच्चय) of equal (सम)/(corresponding) heights (उच्चयः) and the (समुच्चय) as full affine state so attained to be further stripped off its internal multiplicative attributes, degrees and powers, and thereby to have a sequential progression of never ending steps and the end fruit is to be fully transcendental self referral base.
2. This sequential progression attainments is of the features of transcending mind creating

intelligence field for its sustenance in terms of the solid quantifiers to be supplied for the intelligence field by its consciousness field which is to be of the order of the glimpsing of the inner and the inner most fold of the transcendental worlds within Creator's space.

13. GANITA UPSUTRA-13

13.1 TEXT

गुणितसमुच्चयः समुच्चयगुणितः

Gunitasamuccayah Samuccayagunitah.

13.2 SEQUENTIAL PROGRESSION OF STEP: '13'

1	2	3	4	5	6	7	8	9	10
ग्	उ	ण्	इ	त्	अ	स्	अ	म्	उ
11	12	13	14	15	16	17	18	19	20
च्	च्	अ	य्	अ	:	स्	अ	म्	उ
21	22	23	24	25	26	27	28	29	30
च्	च्	ट	य्	अ	ग्	उ	ण्	इ	त्
31	32								
अ	:								

13.3 Functional format

The simple English rendering for the text of this Upsutra may be taken as 'Product samuchya Samuchya Product'. The term 'samuchya' is a technical term, which has many applied values, but the predominant feature of it is that it is to be of the value that lifts upward equally at 'zero state'.

■

SECTION - IV

MATHEMATICS OF ALPHABETS

1
INTRODUCTORY

Learning in terms of ancient wisdom system would require learning of MATHEMATICS OF ALPHABETS.

The self-referral features of ancient wisdom systems bring the beginning and end at the same ment.

With it, the 'alphabet' becomes the 'starting point' of the system as well as the same emerges to be the 'end fruit of the 'knowledge system'.

The 'alphabet' being the starting point of the knowledge system, as well as the same being the fruit of knowledge system', as such, the 'mathematics of alphabets' is of distinguishing features which take care of the 'alphabet' from the phase and stage of settlement of its 'organization format' as well as of its 'functional rules' for making it to be a 'language'.

This focus and attention makes every letter of the alphabet being of a distinct 'format, frame and form'. The features of formats, frames and forms of individual alphabet letters are of such characteristics which accept appropriate coordination arrangements amongst themselves to reach at the full organization format of entire alphabet as to be of the order and potentialities making it capable for attainment of coverage of the entire range of knowledge as a single discipline permitting chase in terms of the emerging language of such alphabet.

With these expectations of systems of knowledge from the formats, frames and forms of the alphabet letters, the ancient wisdom has availed the geometric formats and artifices of numbers for the settlements of these formats, frames and forms of individual letters as well as for the format and functional rules of the entire domain of the alphabet.

Devnagri alphabet is one such classic alphabet of such features and potentialities and because of it having successfully attained the status of the alphabet of language covering the entire range of Vedic knowledge systems. This as such deserves to be chased by thoroughly learning the forms, frames and formats of individual Devnagri alphabet letters as well as about the domain of the Devnagri alphabet itself in its entirety of organization.

There are a good number of other classical alphabets, each one of which, as well, would deserve to be learned and chased for their formats, frames and forms. Of these the one of great functional values is the English alphabet. The glimpse of its different features, may be had from the books (1) glimpses of Vedic mathematics and (2) Vedic mathematics decodes SPACE BOOK. By being through the 'pairing of counts' of the range of artifices 1 to 26 in the sequence and order of alphabet letters A to Z, one may have insight into the richness of the systems of orthodox and classical vocabulary of English alphabet words formulations availing number value formats.

DEVNAGRI ALPHABET FORMAT STEP -1

1. The organization format of Devnagri Alphabet is structurally very rich.
2. Each Alphabet letter has its unique format, frame, form, frequencies flow and it is a properly manifested component of the whole organization format of the Alphabet.
3. The Alphabet letters are accepting multi layered formats which make these letters as (i) Varan (वर्णः), which means that which has its inherent organization and system in terms of which it folds and un folds (ii) Varan (वर्णः) also means colour which makes them the light frequencies (iii) the formulation Varan (वर्णः) is complete in itself and is of unique feature and characteristics for its folding and unfolding inherent organization because of the specific choice and sequence of the letters constituting this formulation in terms of the letters (a) व् (b) अ (c) र् (d) ण् (e) अ (f) ः then (iv) individual letters are also the Akshras (अक्षरः).
4. The letters accept grouping and classification as (i) vowels and consonants. Then (ii) consonants accept classification as Verga consonants and other consonants. This class of "other consonants" permits classification (iii) as Antastha, Ushmana and Yama letters. The Yama letters accept further classification and sub classification.
5. The above settlement of Alphabet letters, as a single group and class, as their first feature, get their location at the middle of the Space and as such these together as a system acquire coverage for the middle portion as such.

6. The system of coverage for the middle has internal characteristic for its extension for complete coverage with additional provision for the beginning and end to make the coverage an exhaustive one. The acceptance of Om for the beginning and Parnavah for the end, that way extends the Alphabet as well as its coverage range and with it naturally the alphabet organization acquires additional features and potentialities as a system as well as an organization.
7. The organization format of Devnagri Alphabet, as such with Om as the beginning and Parnavah as the end and the whole range of fifty letters (9 vowels, 33 consonants and 8 Yamas) for the middle of the space/range, accordingly need be approached as a single complete systems and also as of three folds, the beginning, middle and the end and each of these three folds, as such may be approached independently as well.
8. The independent approach to Om as the beginning part of the Alphabet is to unfold whole range of features for the system.
9. Like wise the independent approach to Parnavah as the end part of the Alphabet, is also to unfold equally very rich range of features for the system.
10. The middle part which is a set of 50 letters when approached yields a very big range of organization characteristics. These organization characteristics are contributed by letters individually as well as collectively for the whole range of fifty letters as well as for the sub ranges of sub classes of the letters as vowels, verga consonants etc. The

artifices of whole number 50 as well in terms of its re-organization contribute parallel organization characteristics for the middle part. Illustratively the re-organization of 50 as $9+16+25=3^2+4^2+5^2$ is parallel to the triple (3,4,5) from whose artifices we may reach the format of right angle triangle of sides of 3,4 and 5 units respectively. And with it, one of the features for the organization format for the middle part emerges to be as that of right angle triangle.

STEP -2

1. These days, the technical approach of computer scripts are taking much liberty with the forms of the Alphabet letters. The form of first Letter Akara of Devnagri script which is availed for Vedic scriptures texts is one illustration on the point. The forms of Letter Akara of different scripts are being inter mixed in the available computer scripts.
2. The first letter Akara is first of the set of Alphabet letters meant for the coverage of the middle part.
3. The middle part coverage is in respect of di-monad format is a coverage of the joint.
4. The joint/middle point/center/origin/knot of pair of parts and the like, well indicate the features of the coverage through the organization of the Alphabet for the coverage of the middle part.
5. The joint of the di-monad/middle of the spatial organization/origin of the spatial order space is a seat of solid order dimensional space and as such the organization format for this coverage/knot/ synthesize of part accepts the need for a solid joint at the joint of di-monad.

6. The form of the first letter Akara, as is preserved in the Vedic scriptures when chased is to take us to its frame and format of Triloky/Vishwa/3-Space as based in creator's Space.
7. With cube as representative regular body of 3-Space and 4-Space as of four Padas/four quarters/four sporting limbs, when chased in the light of the form of the first letter Akara it takes us to the following format, frame and form for this depiction inherent in this letter:

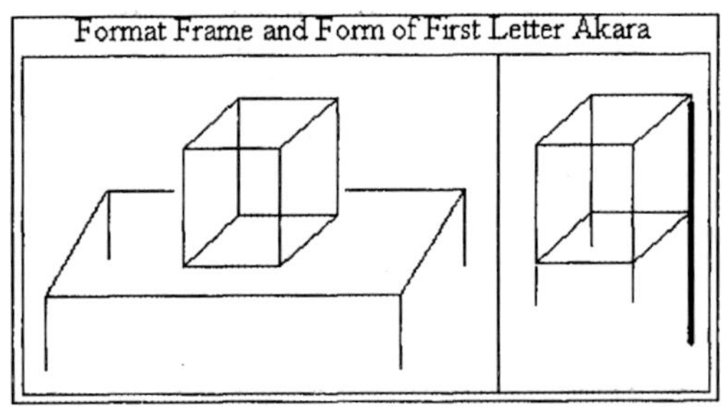

8. With this, this format, frame and form takes us to 3-Space establish at joint of di-monad/center of spatial order/origin of 4-Space (as creator's Space). The 3-Space in the role of dimension for 5-Space and spatial order as dimensional order for creator's Space and di-monad format availed by the spatial order and further the simultaneous happening of transcendence and ascendance through joint of di-monad/center of spatial order/origin of creator's Space, that way also settles flows frequencies manifested as the form of first letter Akara as manifested form with transcendental base.
9. This format, frame, form, frequencies flows

manifestation with transcendental base at the middle/joint of di-monad/center of spatial order/ origin of creator's Space for the first letter Akara, that way also gives us an insight about the organization format with potentialities for covering the joint, synthesizing it for continuity and also chasing the simultaneous happening of transcendence and ascendance at the middle/ joint/center/origin and still further this format, frame, and form also being available as a geometric envelop and carrier for the space content at its middle.

10. This way, the format, frame, form, frequencies flows manifestations, the transcendental base, the carrying capacities as geometric envelops for carrying the space content at the middle of the space for the first letter and also for all the letters which together shall be constituting the desirably full of potentialities as an organization format in terms of which the whole range of reality as a single discipline of knowledge and enlightenment.

11. As such it shall be good intellectual exercise to chase sequentially the format, frame, form and frequencies flows manifestations of all the letters and to perfect the intelligence with respect to the organization format of Vedic Alphabet.

12. Further, it shall be very rewarding exercise to permit the mind to transcend and chase this organization format by riding the frequencies availed by the Vedic Alphabet for its each letter individually and also the whole range of frequencies availed collectively as single integrated Vedic Alphabet format.

FORMAT STEP -3

1. From first to second, is a progression as per the ordering rule of Ganita Sutra-1 Ekadhikena Purvena (एकाधिकेन पूर्वेण:/One more than the before).
2. Here, the second (vowel) following first (vowel), is the ordering as per the rule. However, the ordering also could have been that first the vowels and then the consonants. But in that sequence and order, the whole set of vowels would have been tagged with the first vowel itself and making it the whole range of vowels and then at the second step the shift would have been to the consonants. Then within consonants as well the order could be worked out in terms of the classes of the consonants like verga consonants and non verga consonants. Amongst non verga consonants, it could have been Yamas and others. The others to be Antasthas and Ushmanas. Each of these classes to be sequentially chased as per the orders of letters of these classes. Illustratively Ka (क्) as first consonants, to be followed by Kha ([k) amongst the Verga consonants and out of the five verga rows, to be amongst the first and second letters of the said row. This all is being mentioned just to focus as that while the whole range of Alphabet letters is sequenced and order for the middle part from letters no. 1 to letter no. 50 and simultaneously this range is phased as sub ranges of vowels 1 to 9, verga consonants no. 1 to 25, Antastha consonants no. 1 to 4 Ushmana consonants 1 to 4 and Yama letters 1 to 8.

3. Here, as such, the organization format, frame , form, frequencies flows as are manifesting as second vowel Ekara (इकारःध्इ), as transition and transformation from the first vowel Akara (अकारः/अ) as first component of the organization format of Alphabet, is being approached to have proper insight about the organization format of Alphabet at this phase and stage of this chase.
4. The second vowel is accepting formulation as Ekara (इकारः).
5. Its form is as इ.
6. The phase and stage of first vowel manifests and preserves the initial phase and stage of stitching synthesize for pair of parts of di-monad at its joint as cube as representative regular body of 3-Space with hyper solid-4 base supplied by 4-Space (creator's spatial order space). The center/origin of cube/3-Space as seat of 4-Space/creator's Space presided by creator the supreme, four head Lord, is in a sealed state and as such the center/origin is in un-manifest state. Accordingly for further progression as a system as a step towards stitching synthesize at the joint of di-monad, is to be attained by having process and a step for chiseling to unlock the seal at the center/origin of cube/3-Space, a linear order, to release spatial order for attaining stitching synthesize at the joint of di-monad.
7. It is this processing step as component of the organization format of Alphabet which manifests as form and connected aspect of format, frame as well as frequencies flows of second vowel Ekara

(इकार:) and this focus upon the form of letter Ekara (इकार:), as manifestation of this chiseling processing step as chiseling screw for chiseling at the center/ origin of cube/ 3-Space.

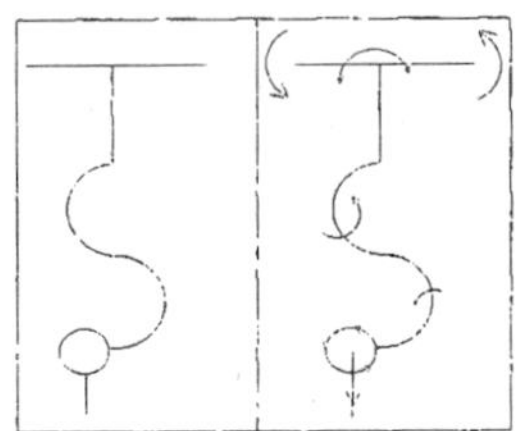

8. The following depiction may help chase this focus for this manifestation and acceptance as a format for the second vowel as organization component of chiseling screw, as the second vowel:
9. The above form of chiseling screw, is further as a chiseling process, has been chased the way chiseling actually is had at the origin of the space to unlock its seal and to have a cut of space as eight octants to manifest as eight components solid boundary for 4-Space during manifestation for its manifested bodies.

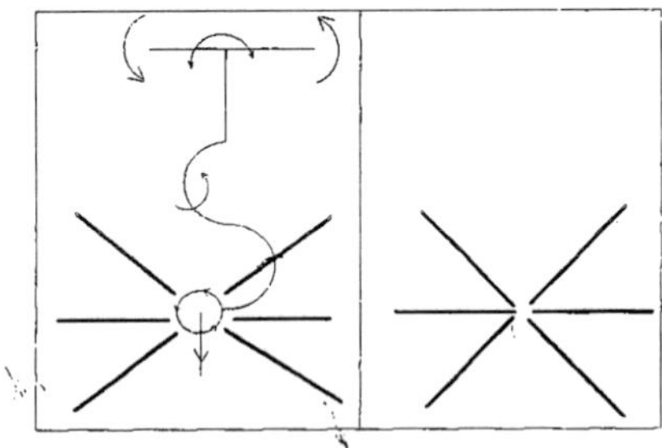

10. This chiseling effect may be further chased as it actually works in respect of solids/cubes as representative bodies of 3-Space with sealed origin which parallel to chiseling of 3-Space chisel out a

space at the center of the cube enveloped within eight sub cubes as eight solid boundary components for hyper cube-4 as manifested regular body of 4-Space.

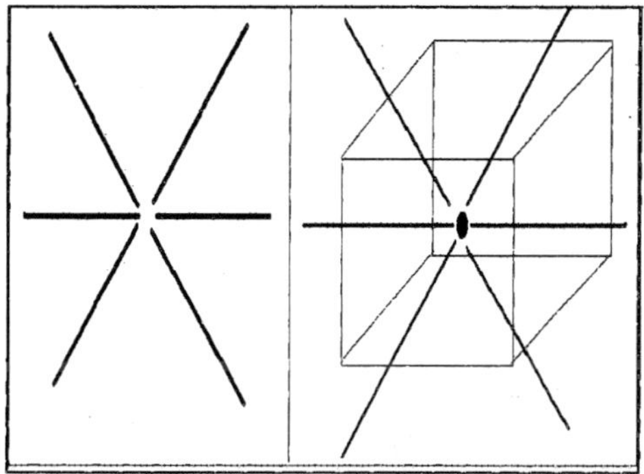

11. It would be good exercise to continue the abbove chase and to reach at the manifested body of 4-Space as hyper cube-4 with domain boundary ratio as A^4:$8B^3$:

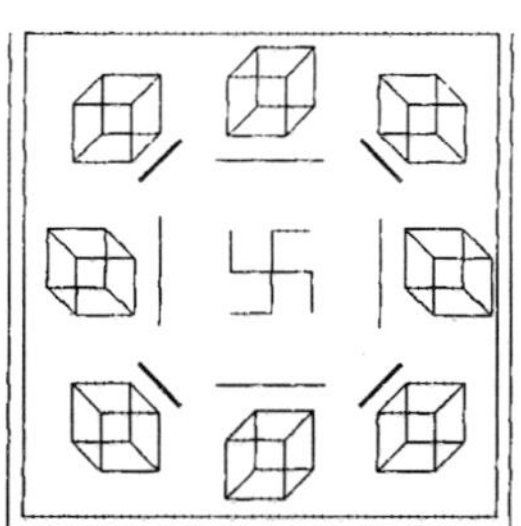

12. It shall be very good exercise to permit the mind to transcend further and glimpse location of seat of 5-Space at the origin of 4-Space/center of hyper cube-4 and to chase chiseling screw processing process to chisel out 5-Space at the origin of 4-Space:

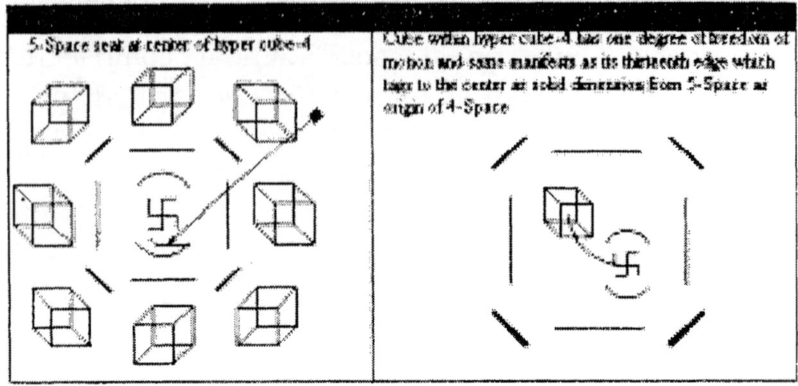

FORMAT STEP -4

1. Before a further sequential step from the organization phase and stage of second vowel Ekara, is taken, it may be helpful to have re-capitulation of the transition from the stage and phase of first vowel to second vowel and also about the attainment in terms of the manifestation of the second vowel (Ekara). The steps number 6 to 12 of the previous chase step number 24 are being reproduced here under:
2. As further sequential step from phase and stage of second vowel, as third vowel, the organization format achieves a step ahead which takes from seat of origin of 4-Space that is from the seat of 5-Space to 6-Space at its base.
3. The spatial order of 4-Space with di-monad format and its chiseling at its origin in terms of second vowel, makes the emerging solid order at the joint of di-monad/center of hyper cube-4/origin of 4-Space as of a pair of hemispheres, which shall be manifesting the form of third vowel and shall be

unlocking the seal at the origin, leading to transcendental worlds, transforming three dimensional frame (of solid order of 5-Space) into a pair of three dimensional frames of half dimensions, and with it attaining transcendence within the dimension as well and thereby the transcendental phenomena coming into play of simultaneous transcendence happening within the domain as well as within the dimensions thereof.

4. With this the organization format acquires a transcendental feature of transcendence within the eternal folds of the transcendental worlds and it is this attainment which makes the organization format of Devnagri Alphabet as of the order of e transcendental features of Vedic systems.
5. This transcendental features of Devnagri Alphabet makes it of the order of the Vedic systems and hence of the potentialities of the order of Vedic Alphabet for cognized knowledge as Shrutis and as the knowledge manifesting with manifestation of Jyoti flowing through the rays of the Sun.
6. The attainment of this sequential stage of third vowel as the self sustenance of the transcendental world (5-Space) with 6-Space as base amounts to transformation of the linear systems sequentially into, spatial, solid and hyper solid systems.
7. With this attainment and breakthrough, the organization format reaches a stage where from the Maheshwara Sutras, Saraswati Mantras and other Vedic systems take over.
8. The organization of knowledge, its systems and the organization format of Alphabet start running

parallel. The expanded expression formats of the grammar availing ten phases of time line (as ten Lakaras लकार:) and ten Ganas of Dhatus/verbs roots, the division of three fold person, three fold genders, three fold subjects and six fold Kaaraks and all other connected aspects of grammar, organizations, systems become unison and with it also become unison the geometric formats, artifices of numbers, manifested frequencies, transcendental bases and the transcendental basis with self referral transcendence for everything.

9. The chiseling at the joint/center/origin of cube, (as representative regular body of 3-Space establish within 4-Space after its emergence from the transcendental worlds and manifestation within 4-Space), takes us to the central double facet spatial plane with three dimensional frame of half dimensions on its either side sustaining the manifestation for the pair of hemispheres, and the spatial order in between responsible for the pair of layers as pair of facet of the central spatial plane becomes the source foundation plane for the whole range of super structure of Vedic systems (including organization of Alphabet) and the organization of knowledge as a distinct but parallel aspect of pure and applied values of the knowledge.

10. As such the organization of Alphabet, Vedic systems and knowledge deserves to be chased starting with the source foundation plane at the joint of di-monad/center of hyper cube-4/origin of 4-Space wherefrom solid order emanates and manifests as a thirteen edged cube with thirteenth edge tagged with the origin and this manifestation

coverage along all the four dimension of creator's Space (4-Space) manifesting the organization format for Vedic Alphabet as of 13x4=52 organization components. It would be a good exercise to intellectually chase the manifestation of Alphabet format of fifty two organization components within creator's Space and also to glimpse and experience and also to chase the emanation of the solid order from within the transcendental worlds and its manifestation within the creator's Space as organization format of Vedic Alphabet with double facet source foundation plane.

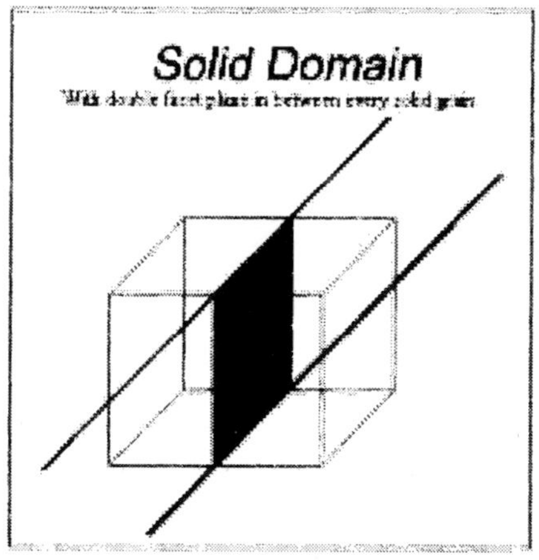

FORMAT STEP -5

1. The organizational chase of Vedic Alphabet, Vedic systems may begin with the central double facet source foundation plane within cube at its center.
2. The double digit expression for numeral for ten place value system with (01, 10) as reflection pairs

of double digit numbers with the feature that the second reflection brings back to the original position and thereby a pair of reflection pairs (01,10), (10, 01) makes the two facet plane as spatial expression of $10 \times 10=100$ grid zones for full expression for the range of hundred double digit numbers starting with 00 and reaching up till 99.

3. From the origin of each of hundred grid zones, is to emanate solid order with emanations from either facet of the spatial zone and thereby would manifest 100x10=1000 cubes on either side of the central plane.
4. Simultaneously the phenomena of emanation of cube at the center of the spatial zone, as a tri-monad format, for each of the three parts shall be constituting $3\times3\times3=29$ cubes and with it the manifestations would get potentialized for recycling as much as that 3-Space in the role of dimension is to structure out 5-Space and 3-Space itself being a dimensional order space, as such there shall emerge transcendence and ascendance within domain (5-Space) as well as within dimension (3-Space).
5. The ten phases of time line (Lakaras) and ten Ganas of Dhatu/verb roots as of the order of one hundred grid zones for the either facet of central spatial plane, and the solid order as 3x3x3 cube parallel to three folds of Pursha, three folds of Gender and three folds of Subjects (Karta, singular, dual and plural) with 1000 cubes expression along each facet of the central plane and the solid order

cube being enveloped within six surfaces, parallel to six Kaarkas shall be manifesting the needed expansion format of grammar, as well as for 1000 branches of Sama Veda.

6. As such, it would be good exercise to first approach the organization of Alphabet, systems and knowledge as geometric formats with central plane as foundation.
7. It would be further good exercise to chase different organization components of expanded expression format availed by the grammar parallel to the geometric formats.
8. Still further it would be very good exercise to chase the geometric formats with central plane as foundation plane parallel to the expanded expression format availed by grammar and to reach at different processing steps of grammar and other Vedangas and other Vedic systems.
9. The intellectual and experiential exercises shall chase the organization format of Alphabet, systems and knowledge with transcendental base and basis sustained by self referral systems of 6-Space/orb of the Sun
10. The touch stone for evaluation of the success of comprehensions of Vedic systems is whether the whole range of knowledge emerges to be a single discipline of knowledge accepting single alphabet.

FORMAT STEP -6

1. The whole range of Vedic scripture are available whose organization formats may be availed for having insight and comprehension of the

transcendental feature of organization format of Alphabet, systems and knowledge.

2. Shrimad Bhagwad Geeta is one such scripture whose organization format may be availed for such insight, comprehension and perfection of intelligence about this aspect as well.
3. The study zone of Shrimad Bhagwad Geeta is 6-Space domain enveloped within the transcendental boundary of 5-Space further enveloped within the manifested format of creator's Space (4-Space).
4. The organizational range of Shrimad Bhagwad Geeta is of 700 Shalokas range of which the starting point is of 47 Shalokas of first chapter and end point is of 78 Shalokas and the middle part of the range being of 575 Shalokas range.

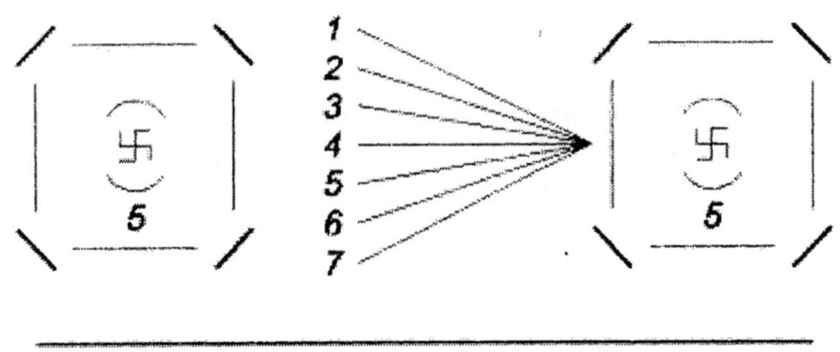

5. The chase of this pilgrimage beginning with the origin of our Vishwa (4-Space) on chariot of Sun driven by seven horses (spectrum of seven frequencies as Ativahkas) and with unlocking of the seal at the origin chariot enters within the transcendental worlds and after transcendental

glimpse pilgrimage completes with return of the chariot. These pilgrimage phases may be chased by availing following depictions.

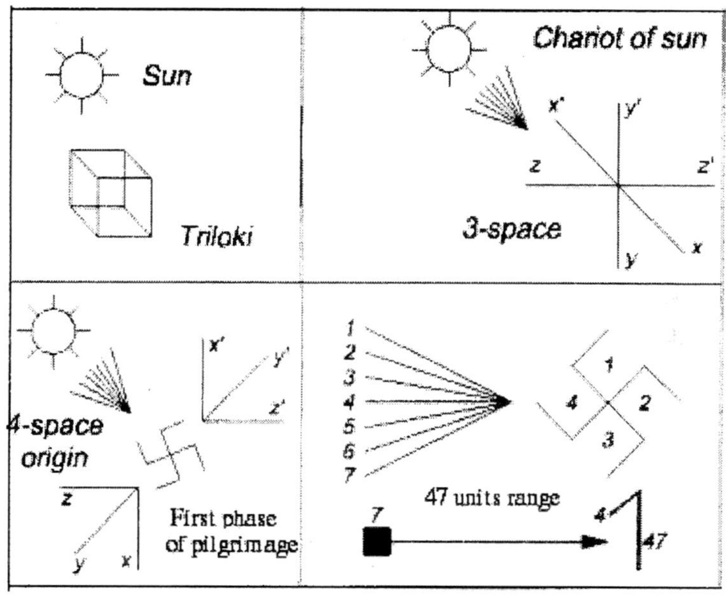

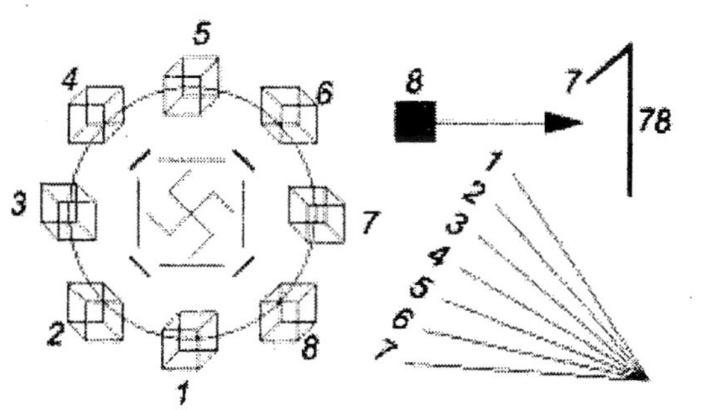

6. It would be very good intellectual and experiential

exercise to chase the organization format of Shrimad Bhagwad Geeta by availing above depiction and glimpsing the completion of the pilgrimage by transcending at the origin, glimpsing inner folds of transcendental world and ascending back to the creator's world and getting ready for the onward pilgrimage.

7. The sequential chase of middle part as of sixteen phases of total 575 Shaloka ranges is very blissful and one shall transcend time and again along the artifices of the Shalokas ranges of Shrimad Bhagwad Geeta starting with 47 Shalokas ranges of first chapter and reaching up till 78 Shalokas range of last chapter.

STEP -7

1. The insight about the whole range of knowledge as single discipline and that the same set of systems working out the organization formats of different scriptures may be well appreciated by simultaneously being through the organization format of Shrimad Bhagwad Geeta and Sama Veda.
2. The organization formats availed are the same and the processing flow lines as well are the same but with the only difference as that in case of Shrimad Bhagwad Geeta the Divya Ganga flow artifices along four components of Om formulation are being availed as transcendence into creator's Space while in case of Sama Veda this very artifices are being availed as ascendance through the creator's Space.
3. The organization format of transcendence through

Om formulation as Divya Ganga flow initiating as seven streams from Bindu Sarovar settles the organization format for Shrimad Bhagwad Geeta. Here below is reproduced the initial flow depiction for three steps as: (the other steps to be chased subsequently).

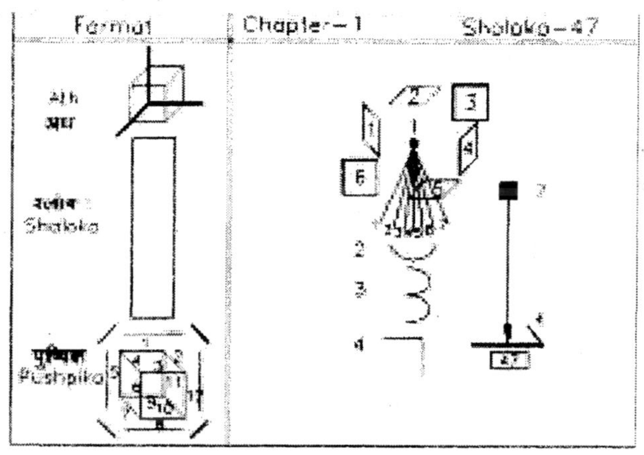

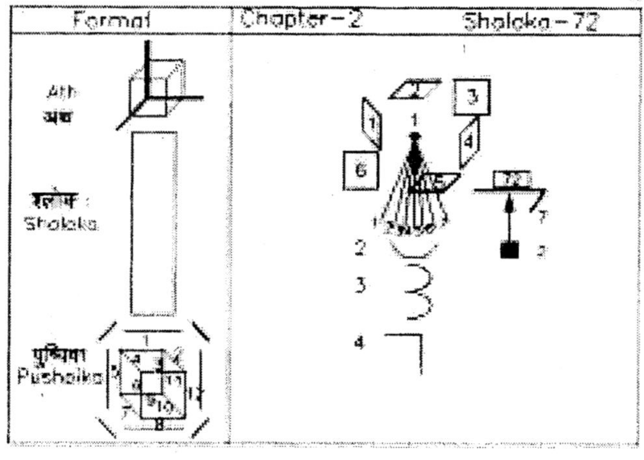

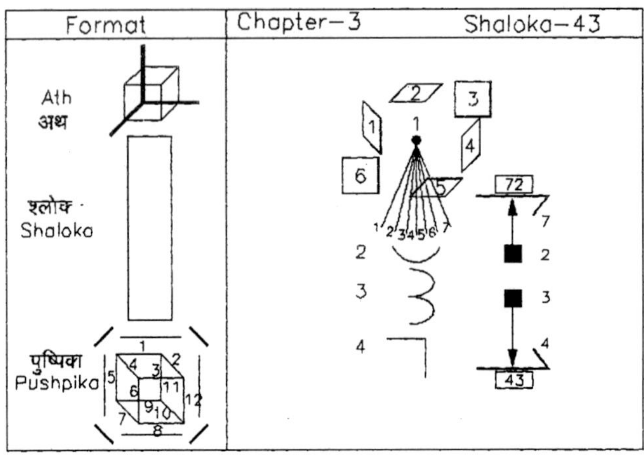

4. The organization format of ascendance through Om formulation as Divya Ganga flow manifesting as Sama Veda, for its initial flow stages as depicted already in the previous lessons of Course-1, is being reproduced below for ready reference:

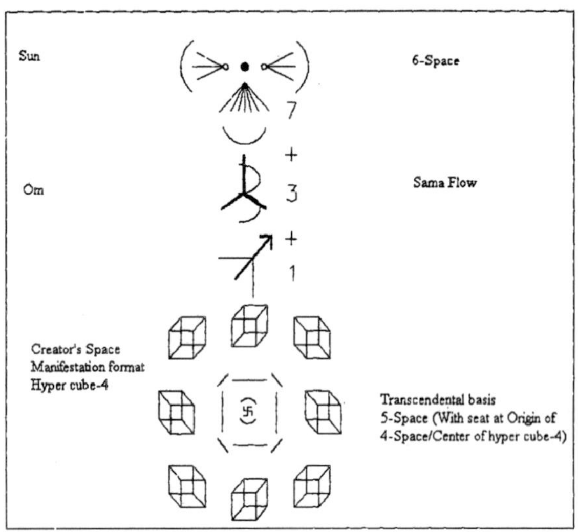

FORMAT STEP -8

1. Basic feature of organization of Vedic knowledge is that it has been organized as a single discipline.
2. The whole range of knowledge has been organized as a single discipline with single alphabet.
3. This feature is parallel to the feature of whole range of existence within single space.
4. Vedic comprehension of this space, as fifth element of Panch Mahabhut, is the transcendental world.
5. Space, as transcendental world of Vedic comprehension is real 5-Space of solid dimensional order with Lord Shiv as the presiding deity.
6. Lord Shiv, the presiding deity of transcendental world, is five head lord with each head equipped with three eyes. The idol of Lord Shiv manifests the representative regular body of 5-Space.
7. The idol of Lord Shiv, five head Lord with ten beautiful arms and each head equipped with three eyes, as phallax and as Murti glorify the features of transcendental world reflected in the manifested representative regular body of 5-Space, which may be designated as hyper cube-5.
8. The unification of knowledge on format of hyper cube-5 as format of manifestation of idol of Lord Shiv, the presiding deity, with five heads and each head equipped with three eyes is the focus of the basic feature which deserves to be chased parallel to the features of the transcendental world prominently reflected in the format and idol of Lord Shiv.
9. Starting with the prominent features of the format

of idol of Lord Shiv and all knowledge and traditions connected there with, the approach focus of geometric formats of Sanskrit Grammar may be chased.

10. There may be different ways to enlist the prominent features of the format of idol of Lord Shiv and the one with geometric focus is to approach it as hyper cube-5, as representative regular body of 5-Space.

11. The enlistment of different locations for reaching at 5-Space, the different roles which 5-Space, its bodies and features play and the different geometrical and mathematical properties of 5-Space and its bodies, are going to be of great help for chase of Vedic knowledge as a single discipline organized availing the format of hyper cube-5 as representative regular body of 5-Space.

12. The Vedic literature focusing upon organization of knowledge in particular and literature of Shaive cult like Shri Shri Shiv Mahapuran may emerge to be the start with source and the same ultimately would take to Vedangas and the apex Vedas. The organization of Vyakarna (Vedanga) accepting Maheshwara Sutras as source formats, as well in furtherance of acceptance of the basic feature as that Vedic knowledge has been organized as single discipline availing formats of the transcendental world presided by Lord Shiv. (For further studies of the topic one may download lessons from website www.learn-and-teach-vedic-mathematics.com .

ENGLISH ALPHABET

Special focus aspect: Unit to unity

Creator's space (4-space / B) is a spatial order space. Being a spatial order space, it has 2-space in the role of dimension. With it a pair of axes are available at dimensional level itself.

The availability of a pair of axes at dimensional level itself, shall be equipping the coordination arrangements within such a dimensional order which are to be of distinct organization format than that of the linear order set up of a 3-space with a cube as its representative regular body.

This distinction may well be chased in terms of the distinguishing features of the set ups of 'cube' and 'hyper cube-4' respectively being the representative regular bodies of 3-space and 4-space.

Viewing from 3-space, the most glaring difference between the set ups of 'cube' and 'hyper cube-4' would be coming within comprehension range of 'sensory field', as that, the pre-dominant expression of 'cube A' is of its volume / solid / domain fold, while, in the context, that of hyper cube-4 B is to be as of 'origin fold'.

However, while viewing from '4-space', the distinguishing feature of a cube from that of hyper cube-4, the dominating expression would be as that the cube is reaching up till the boundary of hyper cube-4.

As such, when the set ups of cube and hyper cube-4 are to be taken up for their comparative chase, the same, as 'domain folds' shall be leading us to the dimensional order features, as much as that the cube, as a domain fold is a set up of linear order while the hyper cube-4, as a domain fold, is a set up of a spatial order.

With this, the macro state for cube as is its domain fold, the same is of a linear order. On the other hand, the macro state for hyper cube 4, as is its domain fold, the same is of a spatial order.

In the light of this predominant distinguishing features of a cube and of hyper cube 4, and that too at macro state level, it may be a blissful chase to follow these set ups, as such, in terms of their dimensional orders, which in case of a cube being of linear features and the same in case of hyper cube-4 being of spatial features, as such the whole focus, naturally, that way gets centered at the properties and characteristics which distinguish spatial order from that of linear order.

The spatial order, predominantly, with the privilege of having a pair of axes, the same, naturally has an advantage over the linear order which is to be satisfied with only a single axis. This advantage of a pair of axes over a single axis, amongst other features, is to be permitting its chase in terms of artifices as being of 'count' to for 'spatial order at count 1' for a linear order. Further, the same, as an operation, the spatial order is to be approached as 'PAIRING'.

The shift from artifice '1' to artifice '2', and consequential shift from 'count' to 'count as pairing', becomes the basic aspect of the ordering system which deserves to be glimpsed, comprehended and chased.

This, when viewed from 3-space, as a linear order, the same is to emerge as of the range from 'cube' to cube end as 'cuboid'. However, the same when is to be viewed from the spatial order point of view, it is to take us to the range of artifice 6, being of the features 3+3. The other features of artifice 6 being as that 6=3+3=2+2+2=1+2+3=1 x 2 x 3 which makes it a perfect number, the same are there because of the pairing operation of artifices.

One chase path for the pairing operation of artifices of numbers, is that, the count 1 as number 1 manifests for it a single artifice. And the same manifests a pair of artifices for 'count 2 / number 2'. These together sum up 'triple artifices'. This emergence, this way takes to the emerging range as being of a set of six artifices.

This feature of pairing operation, is having its full expression within Creator's space / 4-space and here within this domain there being a dimensional frame of quadruple spatial dimensions, as such there emerges a four fold manifestation format with each fold itself being of a spatial order, and thereby the sequential unfoldment from 'whole' to 'parts', being inherently there, and the same is to be taken as being there because of the 'pairing operation', which takes a pair of sequential steps at a time.

The potentialities of pairing operation are there in its feature of 'taking a pair of sequential steps at a time 'which makes the spatial order system taking care of itself by simultaneously processing availing a pair of axes, of which while one of them accepts increasing '**doubling**' progression, the other on the other hand accept decreasing '**halving**' progression.

The simultaneous handling of 'doubling' as well as 'halving' permits the unit to maintain its internal unity intact, as that 2 x 1/2 =1 x 1=1. It is because of this potentiality of the spatial order (2-space in the role of dimension) of the Creator's space (4-space), that the static cube with availability of a degree of freedom of motion within hyper cube 4, acquires an additional edge, that is 13th edge, with which, cube as along a spatial format, shall be a set up of 'artifice 26', and further the spatial format being there because of the spatial order of creator's space, it shall be providing a manifestation format for the set up of the cube, and the same, accordingly, deserves to be chased as such.

3-SPACE MATHEMATICS, SCIENCES AND TECHNOLOGIES

1. '3-space' with focus upon its linear dimensions makes mathematics, sciences and technologies' of this space being of 'linear features'.
2. Cube as representative regular body of 3-space with 'Volume of a cube' as dominant expression being domain fold of this as manifestation layer makes 'mathematics, sciences and technologies' of this space, being of 'physical content lump' features.
3. From minutest granule to the biggest solid, the

whole range, being of same physical content features as well as of identical linear dimensional axes characteristics, as such within each point of 3-space stands inherently in-built a three dimensional frame so 'the mathematics, sciences and technologies' of this space are of 'macro state' features.

4. As the origin from which the axes of 3 dimensional frame remaining sealed and dormant at 0 value, so 'mathematics, sciences and technologies' of this space essentially remain to be of 'monad nature' of linear quantifiers.
5. The monad nature essentially being of features of 'monad' always coming into play 'as a whole', therefore 'the mathematics, sciences and technologies' of this space are to be of 'positive features', and conceptually, the sequential progression is to be of 'one directional flow' alone.
6. Being of one directional flow alone, as such, its processing potentialities are to be only of 'half range', and other half range is always to remain un covered by its processing systems.
7. The working out of second half of the range in terms of the first half of the range makes the processing of this space 'as of forced symmetries', sacrificing the zero and negative artifices attributes.
8. The working bias of 'forced symmetry' deserves to be compensated failing which the Reality is bound to go guised and its restoration is bound to be accepted as an impossibility.
9. The Reality so processed under forced symmetric

acceptance is to deprive of not only of the existence of four and higher dimensional spaces but also the lower, negative, zero as well as one and two spaces as well have to remain beyond reach.

10. The range of 'two and one' spaces would get disguised underneath the processing steps of first and second of the 3 axes of three-dimensional frame of 3-space, and the processing in respect of one and two space, as such would be of those spaces within 3-space, and not as of their independent existence as 1-space and 2-space respectively.
11. Euclids and Descartes were duped, and all of us, not being conscious of this trap, are also the easy preys of it.
12. The answer and remedy for ensured escape is to first concentrate upon the Creator's space (4-space), and then the cube as representative regular body of 3-space be approached as creation manifesting along the four folds manifestation format supplied by Creator's space.
13. 'The cube' as manifested creation is of four folds, of which 1-space is playing the role of dimension and its content lump is manifesting as the dimension fold.
14. '2-space' is playing the role of boundary and its content lump is manifesting as the boundary fold.
15. '3-space' is playing the role of domain and its content lump is manifesting as the domain fold.
16. '4-space' is playing the role of origin and its content lump is manifesting as the origin fold.

17. A step ahead, the chase is to be of the way the transcendental base of Creator's space fulfills the Creator's space with solid quantifiers, and the cube has its distinct role but as along the transcendence range of five folds, the fifth in sequence being the role of 5-space itself otherwise lively at the base of the Creator's space.

18. As such, the sadkhas fulfilled with an intensity of urge to fully know and completely chase the different roles of '3-space' and hence 'the mathematics, sciences and technologies of 3-space' shall permit their mind to transcend following the way Creator's space is fulfilled with transcendental features and the way creations automatically attain self referral features.

TO CHASE THE MANIFESTATION OF CUBE

1. Vedic mathematics, sciences and technologies of 3-space, essentially are to chase the manifestations of solids and in particular the cubes, spheres and cones, prisms and pyramids with rectangular and polygon formats, and 'this specific shapes and forms features of manifestations of 3-space bodies' as such, as well may be designated as chase of CRYSTALS, DIAMONDS and GEMS, as Macro states.
2. Of these, the beginning shall be had with manifestation of CUBE.
3. The manifestation of cube as a set-up within geometric envelope stitched as of 8 corner points, 12 edges and 6 surfaces, in all, 26 geometric components is the first basic feature, which need be chased.

4. Each of the 8 corner points accepts the role of origin for a three-dimensional frame of half dimensions, and thereby, there is available, as many as 8 such frames, which together, shall be constituting a set of 4 three dimensional frames.
5. The coordination of 8 three dimensional frames of half dimensions as a set of 4 three dimensional frames of full dimensions is a feature which deserves to be chased with a focus as the cube as a representative regular body of 3-space shall be having only 3 dimensions and these three dimensions can supply at the most a set of six half dimensions, while the requirement of all the half dimensions for 8 corner points is of 8 x 3 =24 half dimensions.
6. The split for three dimensional frame, even as of three solid dimensions, shall be making available only a set of three pairs of three dimensional frames of half dimensions. It shall, as such be expecting one another pair of three-dimensional frames, and the same, in the circumstances may be hoped having been supplied by the â•˜spaceâ•™ itself. The supply of a pair of three-dimensional frames of half dimensions of solid order is to be there only from the transcendental worlds (5-space) of solid dimensional order.
7. The split of a three dimensional frame of solid dimensional order into a pair of such dimensional frames is to be there because of the availability of Creatorâ•™s space at the origin.
8. It is this availability of the Creatorâ•™s space at the origin which need be chased along its all the

four spatial dimensions, and it is this chase within spatial dimensions shall be providing a format of a pair of orientations permitting inter change there of, because of which the split of a three dimensional frame into a pair of three dimensional frame shall also be permitting reversal as well as translation of them for their setting along the corner points of the cube.

9. It is this pairing of the corner points and simultaneously insertion of a pair of three dimensional frames of half dimensions obtained simultaneously with split of a three dimensional frame and also reversal translation from them to reach and to be established within so paired corner points of the cube, and thereby, making the set up of the cube ensured for its all the four pair of corner points as a real solid set up.

10. The above feature of stitching of the set up of a cube by pairing corner points and embedding three dimensional frames in all the 8 corner points further ensures the stitching for all the twelve edges of the cube as of di monad format with pair of parts of the edge of the cube to be supplied by two different three dimensional frames of corner points coordinated by the concerned edge.

11. This feature of pairing of 24 half dimensions into twelve edges, further in a sequence pairs twelve edges into six spatial frames and thereby is ensured the required six surface plates for the geometric envelope of the cube.

12. This sequential pairing in a pair of steps from 24 half dimensions to 12 edges to 6 surfaces, in that

sequence and order or reverse thereof taking from six surfaces to 12 edges to 24 half dimensions to 8 three dimensional frame of half dimensions to 4 three dimensional frame of full dimensions are the sequential progressions which need be chased for replicating the Vedic mathematics, sciences and technologies.

13. For this exercise, another feature which need be taken account of is the exhaustive coverage for all the 8 corner points in terms of only 7 of the edges, to be designated as the manifested edges, while the remaining five edges to remain un manifested support.

14. This sequential progression for coordination of 8 corner points in terms of 7 edges, is to be of 3 fold orientations along the 3 dimensions springing out from each corner points. The chase along any of such orientations along any of the dimensions at its end reach at the 8th corner in that sequence and order, naturally would also provide reversal for such progression from first corner to 8th corner into 8th corner to 1st corner. This way, two fold progressions, for both of orientations for each edge would make the synthetic stitching of the geometric envelope along the edges to be of manifestation formats of the order of (-1) space playing the role of dimension for (+1) space, and as such the transition and transformation from macro states of edges to micro states of edges would be available. It is this availability and permissibility of transition and transformation from macro states to microstates for the edges would make â•˜0-spaceâ•™ lively in the role of boundary

fold for such manifestation. The 0-space in its role as of dimension fold for 2-space would inherently coordinate corner points as 0-space bodies with surfaces of the cube as 2-space bodies.

15. Such is the richness of the geometric envelope of the cube and replication of it, naturally, can be expected for supplying parallel richness for the mathematics, sciences and technologies of 3-space, provided the whole approach to the 3-space bodies is the way these avail their manifestation formats.

16. The simple progressions emerging from 2 corner points of an interval, 4 corner points of square and 8 corner points of cube, is just to sway away with it a forced symmetries by working up till half range intervals and covering second half parallel to it. The progression 2^N, for its values N= 0, 1, 2, 3, 4 as manifestation layer (0, 1, 2, 3) is to be of the values 1, 2, 4 and 8. There is a jump over artifices â•˜3â•™ and â•˜5, 6, 7â•™. This jump, would be a jump along third, fifth, sixth and seventh edges coordination of 8 corner points, and parallel to it would be a jump in respect of third, fifth, sixth and seventh geometries of 3-space.

17. The chase of seven edges coordination of a cube, would firstly reverse the orientations from that of first edge to that of third edge, and then further the circular orientation of first three edges would have reversal for it along the last three edges, namely, along fifth, sixth and seventh edges. It is this reversal of orientations firstly at third and secondly at fifth, sixth and seventh edges, which

would go disguised for working only with half of interval. The forcing of symmetry by working with half interval is going to be at such heavy structural cost.

18. The seven geometries of 3-space, accept classification as 3 positive geometries, 3 negative geometries and one zero signature geometry. The neutrality of orientation for zero signature geometry is to cause slip with it being its own negative and there by there being four non positive and also four non negative geometries. The working with half interval, as such, in the context is bound to be at the cost of four out of seven geometries in all of three space.

Pairing to Squaring

More insight into the pairing process at the base of manifestation format may be had by chasing the 'squaring' as a word formulation which accepts number value format for it as of artifice 106.

Conceptually, **number value format** (in short **NVF**) is the expression of pairing of artifices 1 to 26 which in the context of English alphabet amounts to alphabet letters A to Z accepting association of artifices 1 to 26 in that sequence and order, and then the words availing particular alphabet letters to yield combined number value format for the whole formulation of the word as summation of the individual number value formats of the concerned individual letters constituting the word formulations.

This, this way, workout NVF (SQUARING) = 19 + 17 + 21 + 1 + 18 + 9 + 14 + 7 = 106. The artifice 106 accepts reorganization as 53+53. This pairing (53,

53) is number values formats pairing NVF (SQUARING) =NVF (AXIS, AXIS).

In the light of above insight, it may help appreciate as to how linear ordered set up, is to make 'NVF' (LINEAR) = NVF (SOLID); and NVF (SQUARE)= 81= 9 X 9= $3^2 \times 3^2$, a paired pairing which shall be taking us to 4-space being a spatial ordered space of 4 spatial dimensions; a creator's space of pairing operation constituting NVF (LINEAR) = NVF (SOLID) In the context, it also would be relevant to note as that, how blissful is the set up of the Creator's space where NVF (MONAD, MONAD)= NVF (TRI MONAD).

AVAILING RICH RESOURCE OF WORDS FORMULATIONS OF PAIRING OPERATION

The orthodox and classical words formulations is a big resource which deserves to be availed fully. At initial stage, these words formulations may be availed for having insight about the partitions of artifices at the base of NVFs of these words formulations.

The second way to avail these words formulations is to have insight about the conceptual terms like pairing, squaring and so on in terms of their number value format.

Illustratively NVF (PAIRING) = 74 = NVF (CONE, CONE) and NVF (SQUARING)=106 = NVF (AXIS, AXIS). Further NVF (PAIRING)= 74 =NVF (ENGLISH(, which as such makes 'pairing' as the basic operation of these words formulations. The partitions of artifice 74 shall further be giving us insight of internal structural characteristics of the pairing features.

Illustratively 74=1+73 amounts to NVF (PAIRING) = NVF 9 (A FORMAT).

Like that it shall be taking us up till NVF (PAIRING) = NVF (CONE, CONE). Then further NVF (SERIES) = 75 = NVF (A PAIRING).

Then, 74 as one of the components of artifices 75 onwards shall be supplying us, the whole range of external features of PAIRING.

The chase like this shall be perfecting intelligence and the same deserves to be availed. The book Vedic mathematics decodes SPACE BOOK may give further insight into this aspect of Vedic mathematics.

It would be appropriate to impress for intensity for the urge to chase the pairing operation availed by word formulation by taking up the illustration of NVF (PERFECTION) = 111 = NVF (CONE, CONE, CONE) = NVF (TETRA MONAD).

It would be blissful for the sadkhas to perfect their intelligence by chasing the pairing operation availed by the words formulations, beginning as:

NVF (PAIRING) = NVF (CONE, CONE)

NVF (MONAD, MONAD) = NVF (TRI MONAD)

NVF (PERFECTION) = NVF (CONE, CONE, CONE)

= NVF (TETRA MONAD)

NVF (UNIT) = NVF (ZERO), a situation where 'zero' and unit are of 'same' format. One may evaluate and indexed ones perfection in terms of ones comprehension and appreciation of 0 and 1 being of same format and unity range.

The idea of the number value format availed by orthodox and classcial english vocabulary may be had from the following illustration:

* **The matter of pages 135 to 138 to continue**

The idea of the wonders of number value formats availed by orthodox and classical English vocabulary may be had from the following illustrations.

Some specific features of space book

1. Space book		NVF(SPACE BOOK)=87
	=	NVF(SPACE PLAN)
	=	NVF(SPACE FRAME)
	=	NVF(TRUTH)
2. Fore-word		NVF(FORE-WORD)=104
	=	NVF(FOUR SPACE)
3. Pre-face		NVF(PRE-FACE)=54
	=	NVF(SUN)
4. Subject		NVF(SUBJECT)=80
	=	NVF(CREATOR)
5. Paper		NVF(PAPER)=56
	=	NVF(LIGHT)
6. Pen		NVF(PEN)=35
	=	NVF(EYE)
7. Ink		NVF(INK)=34
	=	NVF(CONE)
	=	NVF(DARK)
	=	NVF(DUDE)
8. Author		NVF(AUTHOR)=83
	=	NVF(FOLLOW)
	=	NVF(BLACK SUN)
	=	NVF(DARK LORD)
	=	NVF(ONE LORD)
9. Contents		NVF(CONTENTS)=110

	= NVF(MIRRORS)
10. Sections	NVF(SECTIONS)=104
	= NVF(EXISTENCE)
	= FOUR SPACE
11. Chapters	NVF(CHAPTERS)=90
	= NVF(ARTIFICES)
	= NVF(SPHERES)
	= NVF(NEW TREE)
12. Lessons	NVF(LESSONS)=103
	= NVF(COUNTING)
	= NVF(STUDENT)
	= NVF(HEAVEN TREE)
13. Object	NVF(OBJECT)=55
	= NVF(HEAVEN)
	= NVF(SKY)
14. Theme	NVF(THEME)=51
	= NVF(FULL)
15. Aim	NVF(AIM)=23
	= NVF(END)
16. Goal	NVF(GOAL)=38
	= NVF(ION)
	= NVF(FIRE)
17. Conclusions	NVF(CONCLUSIONS)=144
	= NVF(SPACE DISCIPLINE)
	= NVF(HYPER ORIGIN)
18. Virtue	NVF(VIRTUE)=95
	= NVF(RENEWING)

19. Bliss NVF(BLISS)=61
= NVF(CHURCH)

20. Fruit NVF(FRUIT)=74=
NVF(PAIRING)

21. Meditate NVF(MEDITATE)=77
= NVF(CHRIST)

22. Static NVF(STATIC)=72
= NVF(ORIGIN)

23. Moving NVF(MOVING)=80
= NVF(CREATOR)

24. Beginning NVF(BEGINNING)=84
= NVF(COLOUR)
= NVF(GOD TWO)

25. Gods NVF(GODS)=45
= NVF(RANGE)

26. God One, Two, Three, Four, Five, Six
NVF(GOD ONE)=60=NVF(FOUR)
NVF(GOD TWO)=84=NVF(COLOUR)
NVF(GOD THREE)=92=NVF(REVERSE)
NVF(GOD FOUR)=86=NVF(NEW SPACE)
NVF(GOD FIVE)=68=NVF(JOINT)
NVF(GOD SIX)=78=NVF(AMBROSIA)

FOCUS UPON SOME FEATURES

The space plan truth features are reflected in the groups of words-formulations accepting common number value formats. Here below are being focused some of these sub-sets of formulations :

1. NVF(SPHERE)+NVF(SPHERE)

= 71+71=14 =102+40

= NVF(TWO SPACE LINE) =NVF(STRAIGHT LINE)

2. NVF(ORIGIN)+NVF(ORIGIN)

   = 72+72=144=100+44

   = NVF(SPACE DISCIPLINE)=NVF(INTERVAL FRAME)

3. NVF(MONAD)+NVF(MONAD)

   = 47+47=94=NVF(TRI-MONAD)

4. NVF(CONE)+NVF(CONE)=37+37=74

   = NVF(ENGLISH)=NVF(PAIRING)

   = NVF(POINT) =NVF(FRUIT)

5. NVF(BLACK)+NVF(BIBLE)=29+30=59

   = NVF(LINEAR)=NVF(DOUBLE)=NVF(SOLID)

6. NVF(MIND)+NVF(SEED)=40+33=73

   = NVF(FORMAT)=NVF(SOUND)=NVF(PULSE)

7. NVF(MIND)+NVF(SEAL)=40+37=77

   = NVF(MATTER)=NVF(MEDIATE)=NVF(CHRIST)

8. NVF(TRANSCENDENTAL)=NVF(SELF-REFERRAL)

TRANSITION LEAD

The space plan is of such transcendental features for which range, the sadkhas shall go in trans time and again and glimpse the inner folds of the transcendental world and to perfect ones intelligence to chase their transcendental values.

For Further insight about the richness of mathematics of words formulations, one may chase through the referred two books of the Author namely (1) Glimpses of Vedic mathematics and (2) Vedic mathematics decodes SPACE BOOK. ■